INTRODUCTION TO
YACHT DESIGN

SKIPPER'S LIBRARY

INTRODUCTION TO
YACHT DESIGN

FOR BOAT OWNERS, BUYERS, STUDENTS & NOVICE DESIGNERS

IAN NICOLSON,
C. Eng., F.R.I.N.A., Hon. M.I.I.M.S.

FERNHURST
BOOKS

Reprinted in 2024 by Fernhurst Books Limited

Published in 2022 by Fernhurst Books Limited

© 2022 Fernhurst Books Limited

The Windmill, Mill Lane, Harbury, Leamington Spa, Warwickshire. CV33 9HP, UK
Tel: +44 (0) 1926 337488 I www.fernhurstbooks.com

First edition published as *Understanding Yacht Design* in 2003 by Fernhurst Books

Cover photographs © Barbar Studio for Elan Yachts
Sailplan: Humphreys Yacht Design
Boat: Elan GT6

A catalogue record for this book is available from the British Library
ISBN 978-1-912621-44-6

Designed by Daniel Stephen
Printed in the UK by CPI

CONTENTS

FOREWORD 7

INTRODUCTION 9

PART 1: HOW TO BEGIN   11
1   GOOD-LOOKING CRAFT 12
2   HONING THE SPECIFICATION.................... 14
3   SKETCHING THE LAYOUT & APPEARANCE 17
4   DRAWING A DESIGN................................ 21

PART 2: DESIGNING THE YACHT   27
5   DRAWING THE LINES................................ 28
6   WEIGHT & DISPLACEMENT 35
7   HOW WILL SHE PERFORM?...................... 41
8   SAIL PLAN 45
9   CONSTRUCTION 52
10 ENGINES.. 58
11 THE CABIN OR ACCOMMODATION PLAN 64
12 THE DECK PLAN..................................... 66
13 COSTING A DESIGN 70

PART 3: MAKING ALTERATIONS TO A DESIGN   75
14 CHANGING THE SAIL PLAN....................... 76
15 DESIGNING AN ALTERATION..................... 82

PART 4: TO USE COMPUTER-AIDED DESIGN OR NOT?   85
16 COMPUTER-AIDED DESIGN 86

GLOSSARY 90

BIBLIOGRAPHY...94

FOREWORD

I have known Ian Nicolson since we both wrote for *Yachts & Yachting* magazine in the 1970s. I went on to establish a new yacht design company, Humphreys Yacht Design, but Ian took a different route. He joined Mylnes, the oldest yacht design firm in continuous production, in due course becoming the senior partner and owner.

But designing yachts is only one of Ian's areas of expertise. He has also lectured at three universities, written over 25 books, many of which are still in print. Today, aged over 90, he continues to work as a yacht surveyor and race a Sonar regularly at the Royal Northern & Clyde Yacht Club.

It is that level of experience and knowledge about boats that makes Ian the ideal person to introduce you to yacht design. There can be no denying that it is a complicated business, but Ian explains the principles in a straightforward and understandable way.

I have no hesitation in recommending this book to anyone who is interested in understanding a bit more about yacht design, be they a boat owner, buyer or someone thinking about training to be a designer.

Rob Humphreys
September 2022

INTRODUCTION
DESIGN MADE EASY

This book is deliberately short and simple. Chapters have been reduced to basics and liberties have been taken, so that anyone can understand or start designing quickly and easily. Because designing is such a big subject, many problems have been eliminated by using simple 'rules of thumb'. Subjects like stability, speed and powering have been made understandable by substantial simplification.

This book will enable a boat owner to be aware of the characteristics of their design, and it will help them if they want to modify the hull, rig, deck layout or accommodation.

It will be equally useful to boat buyers who want to ask questions like: 'How will she sail?'; 'How much will she cost?'; 'How easy is it to change things?' and 'How can I reduce or increase sail area without upsetting the balance?'.

Boat owners and buyers will be able to feel more confident when talking to designers and boatbuilders about their current or future boat.

While novice boat designers can start their journey with this book, giving them a good foundation whether they want to use pencil and paper or Computer-Aided Design.

A yacht designed using this book should look prettier than some of the products of boat factories. She could well be faster and she should have that most valuable of all characteristics – she will have been designed lovingly by someone who does not constantly seek the cheapest way of making each component. Designs which are heavily influenced by accountants are seldom a success. If a boat is ugly she will not get the love, care and maintenance she needs. She is unlikely to last long or give her owners and crew a full measure of pleasure. The way to learn how to design beautiful craft is to look at the work of the designers whose work has lasted.

Ian Nicolson
September 2022

PART 1
HOW TO BEGIN

A yacht must be curvaceous to be beautiful. This shapeliness has a practical value because it adds to the hull strength, just as a thin eggshell gains strength from its curves.

1
GOOD-LOOKING CRAFT

There are rules for producing a good-looking craft which should seldom be broken:

1 With few exceptions the above-water lines of the hull should be curvaceous

If there must be a flat line, then at least make it short – and the lines at right angles should not be flat – so that there are no flat areas. There is a practical reason here: a curved plate has 'shape strength', whereas a flat one is easily deformed. To understand this, take hold of a piece of paper and wave it. It is flabby and weak. Now wrap it into a tube, and it instantly gains 'shape strength'.

2 Curves should please the eye, and the eye is quickly bored

So lines should alter their curvature along their length. There are certain exceptions to this: a stem which is a straight line is acceptable, a transom seen from dead abeam or the outlines of the keel may be straight.

But almost all other lines of a hull should be curvaceous.

3 The sheer, that is the sweep of the deck seen in elevation, should be a bold curve with plenty of shape

No portion of it should be straight, not even for a short length.

4 The stem and the stern should not rake at the same angle, nor have the same height

5 Lines which are in the same area of the structure above the sheer should have the same slope

So the aft face of the aft cockpit coaming should have the same rake as the aft edge of the cabin top, and aft side of the aft window. If this angle is almost upright, then the fore end of the deckhouse should be much more sloped because 'the eye is easily bored' and does not like forward and aft slopes to be identical or nearly so.

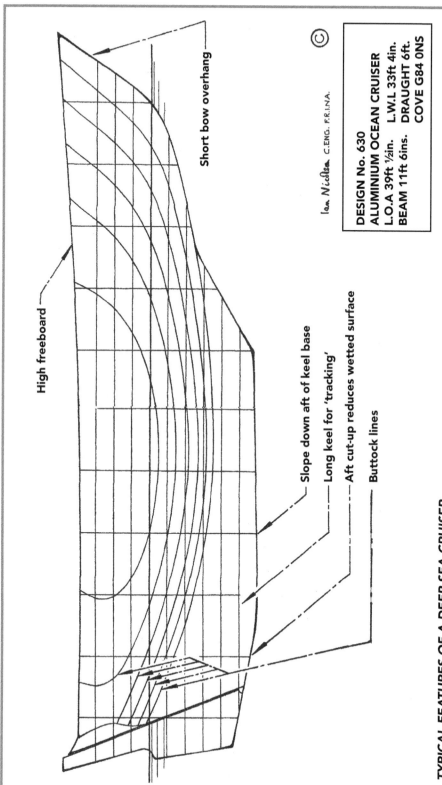

Short bow overhang

High freeboard

Slope down aft of keel base

Long keel for 'tracking'

Aft cut-up reduces wetted surface

Buttock lines

Ian Nicolson. C.ENG. F.R.I.N.A.

©

DESIGN No. 630
ALUMINIUM OCEAN CRUISER
L.O.A 39ft ½in.   L.W.L 33ft 4in.
BEAM 11ft 6ins.   DRAUGHT 6ft.
                  COVE G84 0NS

TYPICAL FEATURES OF A DEEP-SEA CRUISER

This lines plan shows typical features on a deep-sea cruiser. The slope down at the aft end of the keel helps to keep the yacht sailing straight in rough conditions, as does the long keel. This reduces the strain on the self-steering gear.
The aft cut-up reduces the area of boat in contact with the water – the wetted surface – and this enhances the light-weather performance. The cut-away aft also helps turning in confined harbours.

2

HONING THE SPECIFICATION

A professional starting a design will write down the preliminary specification, which describes the vessel and most of her gear, in ample but not comprehensive detail. There will be the main dimensions, approximate size of engine, number of berths, cruising or racing area and a note about the cost. This is followed by an area-by-area list of type, size and quality of each part of the yacht. The rig, deck gear, accommodation, electronics and so on will all be mapped out.

The final specification will be different because, during the course of the full design, owners change their minds, certain items may not be obtainable, may have been modified, found to be just too costly, or are found to have been affected by new legislation.

Beginners (and owners about to go to a professional designer) should make out three 'wish lists'. The first includes all **essentials**. This will normally, but not always, cover such things as the size and purpose of the yacht, the type of construction, rig and number of berths, and anything about which there are strong thoughts – these may include quite small components. Experienced seamen want high fiddles, handy locker clasps and cleanable galleys.

The second list includes **favoured features** which can be sacrificed, especially if something better is found. There may not be enough money or space for everything on this list. **Luxuries** are put on the third list.

THE FIRST LIST – ESSENTIALS

Taking as an example an economical ocean cruiser, her first list includes:

1. The overall length is to be about 12 metres (40 feet)
2. The beam is to be moderate to fit the following requirements

3. The vessel is to be capable of long deep-sea voyages
   - She must 'look after' her crew in all conditions
   - She must 'track' well, i.e. be steady on the helm and steer easily in all wind

speeds, using wind-powered and electronic self-steering units

- She must be able to keep up with other yachts on a 'Round-the-World-in-Company' cruise; but she is not a racing yacht

4. Construction must be reliable with good factors of safety

5. She must withstand accidental grounding in moderate conditions, and be safe in the hurly burly of crowded harbours and canals

6. She will normally only need one person on deck for sail handling; this could be a single-hander's yacht

7. The normal crew will be 2, the maximum will be 5

8. Tank capacity must be suitable for long voyages and include a 'holding' or sewage tank

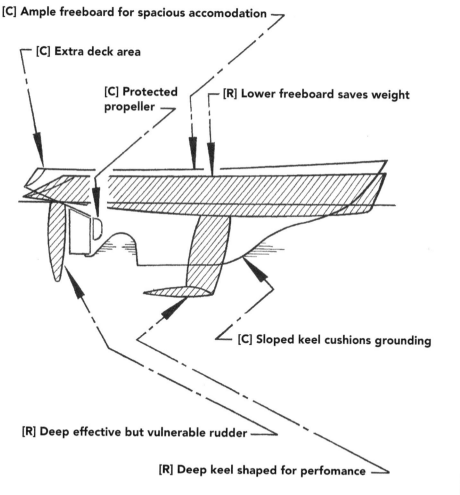

[C] Ample freeboard for spacious accomodation

[C] Extra deck area

[C] Protected propeller

[R] Lower freeboard saves weight

[C] Sloped keel cushions grounding

[R] Deep effective but vulnerable rudder

[R] Deep keel shaped for perfomance

CHARACTERISTICS OF CRUISING AND RACING YACHTS

When designing a racing boat few things matter except speed. Ease of construction, comfort, dryness and many other things are sacrificed.

Cruisers are usually more of a compromise. Adequate rather than maximum speed has to be blended with a moderate cost, space inside and on deck, dryness in the cabin and on deck.

THE SECOND LIST – WOULD LOVE TO HAVE

1. High freeboard and 'flush' deck for a dry spacious deck and ample internal space, also for reserve stability
2. Water ballast tanks port and starboard each holding about 400 litres (90 gallons), as well as a normal ballast keel
3. Tiller steering, for economy and reliability, and for 'feel'
4. A protected rudder and propeller
5. A truly accessible engine
6. An extra large oilskin and 'wet' locker
7. Space for sleeping in the cockpit
8. A workshop, even if tiny, and so on...

THE THIRD LIST – YOUR LUXURIES

The third list may contain anything from a rope cutter on the propeller shaft to a stainless steel tank for whisky in bulk.

Everyone's ideas will vary, and one person will put things on the first list which will appear on another owner's third list.

SOURCES OF INFORMATION

Designers need information, books and leaflets in substantial quantities – see the Bibliography. Browsing through yachting magazines will give the addresses of some manufacturers. The internet is an even better way to find out who, where and what the costs are.

A stroll around a boat show is a chance to collect information as to the sizes and types of engines, cookers, sinks, winches and so on. Sometimes leaflets are unavailable, so it makes sense to carry a folding rule and a notebook. In marinas there are always opportunities to take measurements, or photographs with two or three rules laid beside the fitting to show its width, height and depth.

3

SKETCHING THE LAYOUT & APPEARANCE

The second stage of a design may be a 'library search' to look at comparable designs. A professional may go straight to an old design and start making notes about how the craft will vary. Beginners should rough out a succession of elevation and cabin plans.

MAKING A START

Some people do this on graph paper to give them a scale. Others prefer free-hand sketching, perhaps ignoring scale, at least at first. Once the drawing is complete they note the length of each compartment, add them up to arrive at the overall length of the boat. Others start with a fixed overall length and work in the accommodation.

If there is a shortage of length there are lots of tricks for saving space. These include putting berths one above another or extending berths under the galley worktop or beneath the chart table. Sometimes a slide-out basin can be put over the WC, and a passage can be shut off to form the toilet compartment. A study of plans in books will show how experienced designers squeeze a lot into small hulls.

There are certain traditions and conventions when drawing a design. The first is that the bow always faces to the right to make comparing drawings easier.

Professionals often start off with a similar yacht and then develop it to suit the slightly different requirements that have been specified. This can be done using an existing design on a computer, or the traditional way which is to lay a piece of tracing paper on top of the comparable boat's plan. The designer then roughly sketches in the alterations needed to produce the new craft. A series of changes, made by rubbing out and resketching, eventually produces a plan which, under experienced hands, often needs little alteration during the drawing of the final plans.

A middle-of-the-road technique is to decide on a scale and cut off a piece of tracing paper large enough to encompass the plan and the side elevation of the yacht so that the finished drawing is roughly

500mm (20 inches) across. The idea here is to end up with a plan which the eye can see 'at a glance'. Some draughtsmen prefer to make their early sketches even smaller, perhaps just 250mm (10 inches) across.

DRAWING THE GRID & THE RATIONALE FOR IT

A waterline for the elevation and centreline for the plan need to be drawn using biro or pen so that these lines will not be erased when a rubber is used. Vertical ink lines are drawn at 1 metre (3 feet 3 inches) scaled intervals because a berth or settee is twice this length, and 1 metre (3 feet 3 inches) makes a good toilet space or galley or navigation area. When drawing to an imperial scale, the interval will be 3 feet 3 inches, which is close to 1 metre.

There is a subsidiary reason for using this interval in that it is ideal for major frames and bulkheads. Regardless of the size of yacht, it is easy to give her sufficient strength by having strong frames or bulkheads this distance apart. For economical building the bulkheads and frames should extend athwartships at right angles to the centreline and not be angled or 'stepped' in plan view. If a design is simple, it should not be too expensive to build.

The plan view is sketched showing a

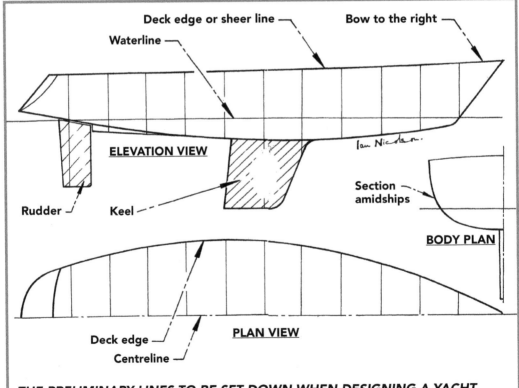

Deck edge or sheer line

Bow to the right

Waterline

ELEVATION VIEW

Ian Nicolson.

Section amidships

Rudder   **Keel**

BODY PLAN

Deck edge

Centreline

PLAN VIEW

THE PRELIMINARY LINES TO BE SET DOWN WHEN DESIGNING A YACHT
This is a typical shape for a modern fast cruiser. One disadvantage of this popular configuration is that damage is likely if the yacht runs aground, especially if she is moving fast. The forward corner of the keel has been rounded slightly to reduce the impact.

typical boat shape, tapered to a point at the bow and a transom at the stern. The beam is taken as average for the size of yacht at this stage. Later it, like the other dimensions, may be altered as the design is developed. The curves from amidships to the bow and stern are made fairly flat at this stage so that the accommodation is not made too ambitious.

One of the commonest faults made by beginners is to cram into the hull the maximum number of berths, lockers and other facilities. This produces an expensive and uncomfortable yacht. A slightly stark interior is the mark of a good design and one which does not cost the earth to build.

SELECTING THE DRAUGHT

Before starting the elevation, the maximum draught of the keel is drawn in the form of a line parallel to the waterline. In the case of an ocean cruiser, the yacht must be able to get into remote harbours, but still go to windward reasonably, and also be adequately stable, so a draught of about 1.67 metres (5ft 6 inches) is often selected.

RATIO OF CANOE BODY TO KEEL

Most sailing yacht hulls are in two parts, the 'canoe body' which is the boat-shaped hull, and the fin keel, which is like a stubby wing extending down from the canoe body. The purpose of the fin is to make it possible for the yacht to sail to windward and it acts something like a sail under water. The depth of the canoe body is:

- About one half of the distance from the deck edge down to the bottom of the keel for a heavy long-range cruiser
- About two fifths for a moderate coastal cruiser with plenty of accommodation
- One third for a boat intended to race as well as cruise. Serious racing craft have shallower hulls

Using this crude criterion, the bottom of the hull is drawn in amidships and this line is extended fore and aft in curves which sweep gently up. The one to the bow normally meets the bow just below the waterline, and the one going aft meets the transom just above the waterline for a counter stern. For a transom stern it meets just below the waterline.

The bow and stern are sketched in to suit the designer. There are a vast number of influences here, but at this stage the aim is to design a pretty yacht. Introducing some science and sea-kindliness comes later.

RULE OF THUMB	
Ratio of canoe body depth: keel depth	
Heavy long range cruiser	50:50
Moderate coastal cruiser	40:60
Racing cruiser	33:66

RATIO OF THE HEIGHT OF THE CABIN SOLE TO BOAT LENGTH

Yet another rough rule of thumb is used to sketch in the cabin sole. For a cruiser it will be about one fiftieth of the boat's length above the bottom of the hull.

HEADROOM

The designer now decides what headroom he wants. These days people are soft and mostly want full headroom at least from the cabin entrance forward to and including the forward cabin. Here it is common to expect enough space fore and aft to undress under full headroom. So a length of at least 600mm (2 feet) fore and aft has to have full headroom.

Just what constitutes 'full headroom' varies according to the height of the person involved, their toughness, experience and a dozen other intangibles. Real full headroom these days is 1.93 metres (6ft 4 inches) clear of the beams and lining. This height is enough for 99% of the population.

On yachts under about 12 metres (40 foot) it is hard to give full headroom right forward, and in plenty of craft 1.7 metres (5ft 7 inches) is accepted. This may be improved slightly at the fore hatch.

The smaller the yacht, the less the headroom and the problem approaches insolubility in craft under 7 metres (23 foot). Various solutions to getting around the problem are used such as large hatches, lifting cabin tops and cabin soles sunk into the fin keel.

THE LID

Now that the cabin sole has been drawn in and the required headroom has been selected, complete with allowance for beam thickness and cabin top deck thickness, the level of the 'lid' on the yacht is known.

This 'lid' is normally a cabin top, but on an ocean cruiser it is often the main deck. In the latter case the sheer, or line of deck at side, is sketched in clear above the required headroom.

At the sketch stage it is essential to make allowance everywhere for future problems. So instead of allowing say 75mm (3 inches) for the beam depth, and 20mm (3/4 inch) for the deck, we assume the two together are 150mm (6 inches) which should give sufficient margin.

4
DRAWING A DESIGN

Boat owners and buyers can skip this chapter about the equipment you need for drawing a design but it will, of course, be useful for those who want to try designing a yacht themselves.

Since it is so important to make beautiful designs, it pays to make lots of freehand sketches of the final appearance of the boat, working these with the preliminary drawings of the cabin plan. Time spent drawing pretty boats and their fittings will subsequently be reflected in the appearance and practical nature of designs.

EQUIPMENT FOR DRAWING

To start designing, a few basic drawing instruments are needed. Naval architects who work mostly on computers still have to make sketches by hand, sometimes at an early stage in the design, sometimes during the building of the boat. The designer also has to put his thoughts on paper when at the builder's premises to explain a detail of construction, or when in discussion with the owner, mast maker or sailmaker.

Even naval architecture students at university do not spend all their time neck deep in computers, especially as there are still some areas where these wonderful tools are less satisfactory than 'traditional' hand work. It is a mistake to plunge into computer work till at least moderate proficiency has been achieved with pencil and paper. For all these reasons some practical hand tools are needed.

Chuck operated pencils which self-sharpen are far better than old-fashioned wood pencils. Two are needed, one loaded with HB leads, the other with 2H. For slick work a pencil with the chuck under the forefinger is best. When the lead breaks or wears, a swift click brings new lead forward and there is hardly a pause in the drawing work.

A set of **pencil compasses** is needed, but these can be of the simplest type provided the hinge is tough and free from any sloppiness.

Soft rubbers are recommended, and a small pliant **brush** is found on professionals' drawing boards for clearing away dust, the accumulation of rubber debris and the crumbs from the coffee break. In this connection it is worth noting that experienced draughtsmen stick down tracing paper to cover over those parts of the drawing not being worked over to keep the job clean.

It is a good idea to have a **drawing board** which is at least 1.5m x 1m (5 feet x 3 feet). Smaller boards can be used, especially for preliminary sketches and minor details, but for much of the time it is convenient to be able to work on one drawing and have another adjacent, spread out for reference.

Anyone who cannot get hold of a drawing board, or who wants to spend less, can buy a flat smooth plywood door from a builders' merchant. The door is secured firmly on top of a domestic table, or on a pair of rigid trestles. The latter are usually best as their height can be made to suit the designer, and the whole affair can be 'triangulated' so that it is firm. One edge of the door is planed absolutely straight for use with a T-square.

Whatever drawing surface is used, and plenty of people have started off on a kitchen table, the surface should be covered with a suitable draughtsman's **plastic film**, sold by the length by suppliers of drawing office equipment. If all else fails, two layers of thick drawing paper are stuck down on the surface.

For holding down drawing paper of all types, heavy-duty Sellotape or **masking tape** is far better than old-fashioned drawing pins. One advantage of working on a drawing board is that paper edges can be held down by Terry's drawing board clips.

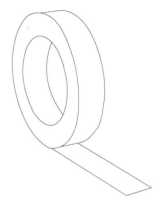

Masking tape

A selection of **scale rules** is essential. Whether they are imperial or metric depends on a lot of factors such as the education of the designer, the location of the boatyard which will build the yacht and current legislation. At present many yards in Europe are prepared to work in either inches or millimetres whereas only a few in America like the idea of metric scales. One of the advantages of working on a computer is that one can switch between scales instantly.

As a rough guide, the hull length on the drawing is best confined to a width of about 700mm, which is approximately 2 feet 3 inches. The eye can encompass the

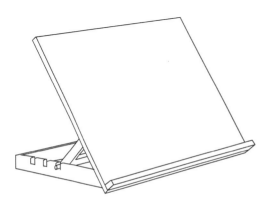

Drawing board

whole plan at once and, unless the vessel is over 14 metres long (around 46 feet), the detail will not be too small.

From a practical point of view it is best to make a drawing large with bold lettering because the people who use it during the construction will be able to see everything clearly even after the print has been creased and soiled.

The ideal scale rule has on it all the different scales being used for the yacht. On one side the scale may have: one-inch-to-the-foot for the lines, construction and cabin plans; with half-inch-to-the-foot for the sail plan or above-water elevation. On the opposite side there will be three-inches-to-the-foot, one-and-a-half-inches-to-the-foot and three-quarters-of-an-inch-to-the-foot, the first two of which will be handy for enlarged details.

A **good light** is essential, and here, as elsewhere in the field of naval architecture, there are as many opinions as practitioners. The first aim must be to have no shadows on the drawing board. This requires something like a 1m (3 foot) fluorescent strip light each side at a height and distance which ensures there is neither a fierce glare, nor shade. Ideally the lights are suspended in such a way that their height can be varied.

A good light

Many people like to have an Anglepoise light as well as the overhead lights. This close-range lamp stands on the drawing table or is clamped to the edge of the board. Its head can be moved in all planes so that the intensity and direction of the illumination can be varied.

An ordinary bulb is not ideal because it gives off too much heat and sometimes too much glare. A modern long-life coiled glass bulb works well. It is, in effect, a strip light made compact. It is available with an ordinary bayonet or screw fitting.

Anyone with surplus money may be tempted to buy some sort of **draughting machine**. These come in a variety of types and sizes and, in essence, consist of a vertical and horizontal scale plus straight edge, articulated so that they can be moved anywhere on the board. They ensure that the draughtsman gets accurate lines where wanted and they speed up general drawing work. They can be adjusted to give parallel angled lines and play all sorts of 'draughting tricks'.

Only the expensive ones are truly accurate, and a worn one must be of doubtful accuracy. For preliminary design work and for fittings they have much to commend them. However many people who want to go beyond basic simple design work will graduate to computer designing, by-passing a draughting machine.

All in all, a good case can be made for sticking to simple traditional gear during the learning stages. A youngster at school or a pensioner can have endless enjoyment with the relatively inexpensive equipment described in this chapter.

For drawing the baselines, waterlines and the like, a long **straight edge** is essential. It has to be longer than the longest straight line to be drawn. When starting as a designer it is hard to predict how long this will be but, if in doubt, go for a minimum length of 1.3m (4 foot 3 inches).

These days there is a selection of equipment like straight edges and T-squares. Some of it is old-fashioned but not necessarily to be despised, some is modern and made from stable plastic which resists denting and stays accurate in spite of changes of temperature.

If a draughtsman's straight edge is hard to get hold of, a plain steel straight edge as used by engineers and house-painters will do, provided it is carefully used and kept rust-free. Alternatively, a beginner can use a straight edge made from a suitable piece of well-seasoned hardwood. Its edge must be carefully planed by an experienced craftsman.

To check the accuracy, lay the edge down and draw a fine line along it. Turn the straight edge end for end and draw a second line. The second line should exactly coincide with the first. If it does not, that is where the edge is not fair. Better still, lay the straight edge along a friend's proven one and check that no daylight shows through between the two edges.

A **T-square** is needed for drawing accurate vertical lines, but it is not essential. Instead the straight edge and a set square can be used. Before buying a T-square, test it by laying it on the edge of a drawing board and draw an accurate fine pencil line down the blade. Now turn the T-square upside down and align it with the edge of the drawing board by using a plastic set square held vertically between the board's edge and the short cross-piece of the T-square. Sight the blade down the pencil line and draw a second line. It should pass exactly down the first line. If it does not the T-square is not accurate.

It pays to buy high quality instruments because cheap ones often hamper good work and take the pleasure out of the job. In this connection the T-square should extend almost the full height of the drawing board even though a small one will be cheaper. Its working edge must be of a material substantially harder than wood.

A minimum of two **set squares** are needed. One will be roughly 170mm (7 inches) and the other 270mm (11 inches) along the shorter edge. These days this sort of equipment is made of transparent plastic and it is sufficiently stable to remain accurate to the standards we need.

The type of set square with bevelled edges is best because it can be used with the bevel on top when drawing a line with precision. When working with a pen, the square is inverted so there is no risk of the ink running underneath. Skilled draughtsmen also use the square this way up when they want to draw a line with a tiny subtle curve. For instance, the shape of the front of a mast tapered above the hounds is almost straight, and an experienced draughtsman can get it right by eye using a bevelled set square.

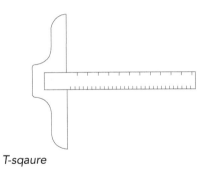

T-sqaure

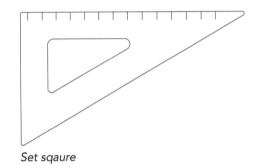

Set sqaure

A naval architect differs from other artists in that he draws lines using the edge of an instrument to guide the pencil. If the line is short and straight he will normally use a set square. However boats of all types have rounded bulbous hulls, so shapes called 'curves' are needed to control the path of the pencil. These curves come in two basic types, though there is a blurring of the division as some seem to fall into either type:

- Long curves, which often look like the blade of a scimitar, having one edge in a gentle sweep and the other a similar but slightly bolder shape.
- Smaller curves with a perimeter which turns more sharply are sometimes called 'Dixon Kemp pears' after their inventor. They look like pears which have grown slightly askew, and they vary in size being typically 80mm up to 200mm in height (roughly 3 to 8 inches).

These can either be bought from suppliers of draughtsman's equipment, or made from thin transparent plastic. If they are being fabricated the outlines can be sketched by hand, or copied from a friend's set of curves. The shape is marked on the plastic and cut out, then smoothed carefully to ensure that there is not even the slightest sign of a flatness or bump.

Purists will not use curves for the important lines of a boat's shape. They insist that only **battens** held down by lead weights can be used. These battens are long thin strips of a semi-rigid material such as lancewood or a transparent plastic like Perspex. Again they may be bought or made. In theory any hard wood which is close-grained as well as totally free from knots and blemishes can be used.

For making plastic battens, Perspex sheet about 3mm (1/8 inch) thick is popular. A batch about 1.3m (4 foot 3inches) long is cut with various tapers. It is best to make six or even ten, because different designers have different needs. After a few months it will probably be found that three or four battens are favourites. The rest are seldom used but may be needed if one of the better ones breaks.

These battens can be used on the flat or on their edge. They are faired off with glass-paper after being cut out, to ensure they are smooth all over. Two are laid close together to check no daylight shows between them – if there is, there is an unfairness.

The thickness and taper of wood battens will vary to give different types of curving line. The sheer of a typical yacht is almost straight, so it is drawn by using a thick stiff batten, perhaps 6 x 5mm (1/4 x 3/16 inch). However, for the sharp bend at the aft end of the waterline, the batten may not bend to the required shape unless it is down to 3 x 1mm (1/8 x 1/16 inch) or less.

To hold battens in place while a line is drawn along their edge, between six and eight 2kg (4lb) **lead weights** are used. These are not easy to buy so many people make their own. They are covered with Fablon or glued-on cloth to prevent the lead soiling the hands.

Not all drawing is done with pencils. Base lines and the grid used for the lines plan must be indelible so that when the pencil lines are altered by being rubbed out, the grid does not also disappear. For these reference lines ordinary biros can be used. The mark made by a fine-pointed biro is thin and sharp-edged. Just as important, biros are available in different colours so the waterlines and level lines will be in blue, and the vertical lines in a contrasting colour. Red is used traditionally but it does not reproduce well when the plan is printed so another colour may be chosen.

Biros are also useful for preliminary sketches because they give a narrow line which can so easily be hardened up by repeated strokes, or left feint. The outline may be done in black, the settees outlined in red and maybe their cushions shaded in the same colour, the woodwork done in brown and so on. Practice and enjoyment will improve the standard and usefulness of the early sketches.

The available choice of **papers** and paper-substitutes increases every year. A beginner should buy a roll of 112gsm (grams/m^2) natural tracing paper because it is tough enough to stand up to rubbing out repeatedly and is not expensive.

It copes with hard usage and, when the lines plan is complete, a new layer of tracing paper is laid on top so that the boat's outline can easily be traced for the start of the construction plan. Later another layer is placed on top for the cabin plan. This technique works because 112gsm paper is reasonably easy to see through.

In time every naval architect assembles a range of equipment, some of which goes out of favour when new devices are invented. In the same way he graduates from basic tracing paper to the latest 'miracle' plastic drawing office film which is stable, easy to use, but may cost four times as much.

Drawing pens are needed for producing final drawings to the best possible standard. They are also needed for setting out baselines when biros are not used, and for putting the neat border round each plan to give it a professional touch.

The rate of improvement in draughtsman's pens has been so fast recently that it is a risk to recommend anything in case it has been superseded. Here again it pays to take the advice of a buyer in a firm of draughtsman's suppliers.

Good quality pens are found in shops selling artists' supplies. If there is no supply of paper in the shop for testing pens before buying, maybe this is not the ideal source of pens. A pair of compasses which hold ink pens can be useful but is seldom essential.

The whole point about ink lines is that they are stable and not easily eliminated. If a mistake is made, a special short-bristle **nylon brush** is needed to erase the ink. These brushes wear quickly, so they have a type of chuck which screws out more bristle length as it is worn away. If fine granules of the nylon get onto the skin they can be painful, which is where the drawing board brush comes into its own.

Normally a **planimeter** is needed to discover the area of each section. This relatively simple precision instrument is not cheap, and if the designer is soon going to change to using a computer, there will be a natural reluctance to buy a planimeter.

Squared graph paper can be used by a beginner, to save money at the cost of a lot of time. Planimeters can be bought second-hand, but they do need careful checking by running the pointer round a known area and comparing this with the reading on the dial.

Various **stencils** are a help, including ones for producing a range of circles, ovals, hexagonals, lettering and so on. The type with bevelled edges are best. Admittedly a great deal of work can be done without stencils but, like other gadgets, they speed up the work and add a touch of professionalism.

In practice, a beginner can make plenty of progress and have a lot of enjoyment with just a pencil, rubber and biro, a straight edge plus two set squares, a few basic curves and a scale, as well as a roll of medium-priced tracing paper.

PART 2: DESIGNING THE YACHT

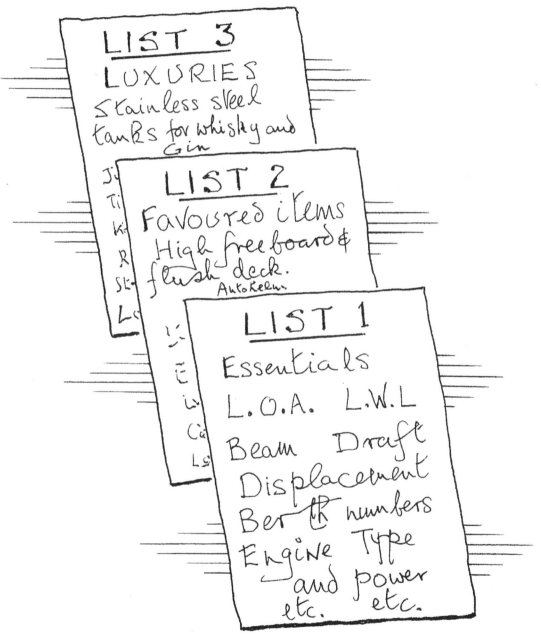

LIST 3
LUXURIES
Stainless steel
tanks for whisky and
Gin

LIST 2
Favoured items
High freeboard &
flush deck.
Auto Keem.

LIST 1
Essentials
L.O.A.   L.W.L
Beam   Draft
Displacement
Berth numbers
Engine Type
and power
etc.   etc.

It's an excellent idea to make three lists of requirements. The first list consists of the obvious features such as the dimensions. The second list is the desirable but non-essential things. The third is the luxuries… if there is money available for them.

5

DRAWING THE LINES

A designer explains to the boatbuilder what the precise shape of the hull is to be by producing a 'set of lines'. Because the boat is curvaceous an ordinary house architect's drawings are useless since they can only define rectangular and triangular shapes or perfect spheres.

THE LINES – WATERLINES, BUTTOCKS, SECTIONS & DIAGONALS

A naval architect draws a set of contours, similar to the contours depicting hills on a map. However hills do not have to be as accurately portrayed as the shape of a hull. So whereas hills are drawn only with contours seen from above (that is in plan view), boat lines include four sets of contours.

The ones seen from above are called the **waterlines**. Sometimes the waterlines above the Designed Water Line (also known as the Flotation Line) are called Level Lines.

The contours seen from the side – always the starboard side – are called **Bow Lines and Buttocks**. These are almost always shortened to simply 'buttocks'. In theory the bow lines extend from the stem back to amidships and the buttocks extend from amidships aft.

The contours seen from forward or aft are the **sections**. They are what would be seen if the yacht was cut downwards with a saw at regular intervals, starting near the stem and working aft. One set of cuts is shown opposite.

The fourth set of contours are the **diagonals** which are roughly halfway between the waterlines and the buttocks and are explained in the caption opposite.

The lines must be drawn with great accuracy and this takes practice. For this reason it is a good idea to start using tracing paper and aim to do the job in a day because tracing paper changes shape with alterations in temperature and humidity. When doing a final lines plan from which a yacht is to be built, a plastic film drawing material should be used. These thin plastic sheets are dimensionally stable and are sold by drawing office suppliers. They are expensive but tough and long-lasting.

Alternatively, the traditional material was a good quality thick cartridge paper.

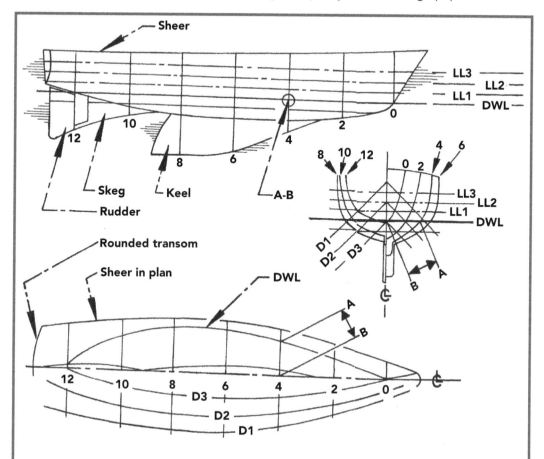

LINES PLAN SHOWING THE BUTTOCKS IN SECTION & THE DIAGONALS IN PLAN

This is a simplified lines plan, using alternate stations numbered from 0 to 12. To keep the drawing uncluttered the odd numbered sections have been left out. When making a preliminary displacement (weight) calculation only the even numbers are used, to hurry the job through.

The width of the waterline at Station 4 is shown as A-B on the plan view at the bottom. This width is the same as A-B on the section view to the right. The same line is a single point on the elevation at the top of the picture, as it extends 'into' the drawing.

This is an essential feature of a lines plan. The distance off the centreline, or the distance from the Datum Water Line (DWL) is shown on two out of three views, and in the third view it is a single point. (Incidentally DWL is sometimes called the Designed Water Line, and it is normally where the boat floats when fully loaded.)

All the fore-and-aft lines must be smoothly curvaceous to ensure that water flows effortlessly over the hull. If the water cannot flow easily, the boat will be slowed down.

Diagonal lines on the section are drawn at 45° for convenience, and labelled D1, D2 & D3. There can be any number of these lines, and they are used to help make the hull shape fair and sweetly curvaceous. The distances from the centreline along each diagonal, at each section on the body plan at the right of the picture, are measured and plotted on the plan view at the bottom.

MAKING THE LINES FLUENT

A good technique is to do a lines plan on tracing paper and pin it up on a wall where it can be studied.

Look at it long and hard. Think about it for a few days, then hide it away, and have another go, eliminating mistakes and setting out to do a more shapely, sleeker boat.

Pin this new lines plan up alongside the first and study them both for a few days. They will probably look similar but soon one will be seen to be better than the other.

A beginner should draw several lines plans of a dreamship before even considering doing a final one for the builder on plastic film. Pinning completed plans up on a wall at eye level for prolonged study is a practice which is recommended for all drawings, not just the lines.

Lines plans should be as big as possible but not so large that the eye cannot encompass the whole drawing at once. Since a designer often wants to compare a new design with a previous one it is logical to stick to the same favourite scales year after year, and not change them without good reason.

CHOICE OF SCALE

For a yacht of about 40 foot long use a 1/16th scale, or 3/4 inch to one foot. Anyone preferring a metric scale will probably go to 1:10 because 1:15 is seldom used and scale rules this size are rarely found. We could select 1:20 but it is too small. This is one of those occasions when using the metric system is awkward.

LAYING OUT THE DESIGN

The piece of tracing paper you cut off the roll should be large enough to show the yacht in elevation (that is side view) at the top, with the plan view (as seen from above) located below. The sections may be drawn on top of the elevation amidships, but a beginner will find it less confusing if they are drawn to one side, as shown on both P18 and P29.

The piece of tracing paper has to be long enough for the sections at one side, and wide enough for the diagonals (which look like a second set of waterlines) on the lower half of the plan view.

The lines plan is drawn on sets of straight lines forming grids, as shown on P29. Each grid has horizontal and vertical straight lines and a set of curved lines, because all the lines are interrelated.

In the **elevation view**, at the top, the horizontal straight lines are the waterlines and level lines, the vertical straight lines are the sections and the curved lines are the bow-and-buttock lines (these have been omitted from P29 but are shown on P13).

In the **plan view** at the bottom the horizontal straight lines are the buttocks, the vertical straight lines are the sections and the curved lines are the waterlines and level lines.

The **section view** has horizontal straight lines which are the waterlines, vertical straight lines which are the buttocks and the curved lines are the sections.

The grids should be drawn in ink using the thinnest lines possible. All the curved lines at this stage are drawn in pencil because they need frequent alterations.

Inked-in grids are in theory unaffected by the rubber and so save a lot of time and potential inaccuracies which would occur if the grid was in pencil and had to be repaired after being rubbed out.

Some energetic draughtsmen, when using rather hard rubbers, have been known to erase parts of an ink-lined grid. To get around this problem a designer may consider drawing the grids on the back of the tracing paper.

The waterlines and buttock lines are typically drawn at intervals equal to about one foot, which in our case scales at 3/4 inch. Some designers tend to have the waterlines from the flotation line downwards more closely spaced than the level lines. This is illogical because, once at sea, the designed waterline, which is also called the Load Water Line (LWL), means very little. The yacht is heeled, pitching, maybe rolling, and the sea level is anywhere but neatly along the LWL.

It is the normal practice to divide the waterline length into ten equal parts. These divisions are numbered 0 at the fore end of the waterline and 10 at the aft end, and they form the sections.

However, there is a good case to be made for dumping tradition and having twelve sections because the designer uses Simpson's Rule to find out the displacement of the hull. This rule is described on P33 and it will be seen that, if there are twelve sections, then every alternate one can be used for a quick preliminary check of the displacement.

Drawing lines is a personal affair, something of an art, so there are different approaches. When gaining experience a good technique is to use an HB pencil and sketch the basic outlines freehand. These are:

1. The sheer and the whole outline of the yacht in elevation from the top of the stem downwards, along the keel, and up to the top of the sternpost
2. The plan view of the deck and load waterline
3. The midships section and a section well forward and another well aft

These first lines establish the size and type of yacht and they must enclose a shape which is large enough to hold the accommodation. It is a mistake at this stage to be parsimonious: the length, the beam, the freeboard, in fact all dimensions, should be at least a little larger than is essential for the job.

When the sketched lines look sweetly curvaceous, and the shapes all look right, they are firmed in, often with a 2H pencil. Normally this will be done with flexible battens held in place by weights. However the sections will curve so sharply that battens cannot be used and ship's curves have to be used instead. These curves can be bought from a draughtsman's suppliers or they can be made from Perspex or another thin plastic sheet a little under 3mm (less than 1/8th inch) thick.

A beginner should compare his lines with examples from a book to ensure that the end product is going to be practical. Once the basic lines detailed above have been drawn, the intermediate ones are plotted.

FAMILIES OF INTERRELATED CURVES & LINES

As an example, take the level line halfway between the LWL and the deck. We know where it begins and ends by checking on the elevation as to where it cuts the bow and stern so these points can be marked on the plan view. From the three sections drawn we know its width at these three points, so we now have five points to join up.

The resulting curve should look like a half cousin to the deck line in plan, and a half cousin to the LWL. That is to say the newly drawn level line should have a 'family resemblance' to the other lines on the plan view. This trait applies equally to the buttock lines and the sections.

Next a waterline halfway between the LWL and the keel is drawn in, as well as a buttock line halfway between the centreline profile and the maximum beam. This gives us four curves in plan view:

- At deck
- Halfway down the freeboard
- At the LWL
- Halfway down to the keel

We also have two curves in elevation plotted in:

- The outline of the yacht
- The single buttock

Using these six lines we have four widths for the next lot of sections. We also know the exact point where the deck edge is for each section since the sheer line gives its height and the deck line gives its width. We know where the bottom of the section is, namely on the centreline and the distance below the waterline shown in elevation.

The distance down one buttock is also known too. It is now possible to draw intermediate sections between the existing ones. However, at this stage, snags start to appear. It may be found that it is impossible to draw a smooth fair curve joining up all the points defining a section. So the next job is to 'pull in' and 'push out'

the waterlines, and adjust the buttock lines till the sections can all be drawn as smooth curves. This is known as '**fairing**'.

Little by little all the buttock lines and the waterlines are added, then alternate sections. At this stage a preliminary displacement calculation needs to be made to ensure the volume of the hull below the waterline is going to be correct (see P33).

To measure the distances off on the lines plan one can use a pair of dividers. However this is slow and the sharp points may pierce the paper. Instead most people use 12mm (1/2 inch) wide strips of paper which are laid along the straight line and the distances marked by pencil ticks. When drawing a level line, for example, the paper is laid on the sections and the width of each section is ticked off, as well as the centreline. These ticks are then transferred to the plan view, section by section, and the points joined to form a new curve.

Diagonals are drawn from the centreline outwards and downwards in section view. Their purpose is to give an extra set of curves to be sure the hull is free from bumps and hollows.

The angle down and the location of the inner end of the diagonals is a matter of choice, and is sometimes selected so that the diagonals cut the outer sections at places where the waterlines are few, or where they are at an oblique angle to the sections and so cross at an imprecise location.

Some designers like to be able to compare one design with several previous ones at different stages, so they always draw their diagonals at 45 degrees, and always start them where the waterlines intersect the centreline in section view.

DISPLACEMENT CALCULATIONS

We need to find the volume of the hull below the Load Water Line (LWL) so we measure the area of each section below the waterline. This can be done in three different ways:

1. With a planimeter, a drawing instrument sold in shops supplying draughting materials.

2. By dividing each section into small right angle triangles and working out the area of each, then adding them together. The area of a right angle triangle is:
   One half of the length x The width
   (These dimensions being along the two short sides.)

3. By laying transparent grid paper, which is divided into little squares of known size, and adding together the number of squares covered by each section.

Where the curved line on the section runs, it will cut through some squares. If more than half of the square is inside the section, that square is included, and if less than half the square, that one is omitted from the count.

This type of paper can be bought, or it can be made up by drawing squares of a required size on a piece of tracing paper. The smaller the squares, the more accurate the calculation will be.

SIMPSON'S RULE

To work out the volume of the hull below the Load Water Line, Simpson's Rule is used. A typical calculation is shown in the Table below, using alternate sections when there are 12 sections.

SECTION N°	AREA (FT²)	MULTIPLIER	PRODUCT
0	0.00	1	0.00
2	0.32	4	1.28
4	0.87	2	1.74
6	1.49	4	5.96
8	1.35	2	2.70
10	0.90	4	3.60
12	0.00	1	0.00
			Total = 15.28

The sum of the products, 15.28, must be multiplied by 2, as only one side of each section has been measured, so the summation is 30.56.

If the distance between alternate sections is 6 feet then the volume of the hull in cubic feet is:

1/3 x 6 x 30.56 = 61.12ft³

As there are 35 cubic feet of salt water per ton we divide 61.12 by 35 to get 1.75 tons displacement. (The figure for fresh water is 36ft³.)

If the boat is going to weigh say 1.9 tons, then the volume of the hull below the waterline has to be increased to give the extra 0.15 tons. This is done by redrawing the sections fuller, then making the waterlines and buttocks conform to the new section shapes.

The reason for using alternate sections for the calculation is to save time. Once the hull shape is right, a final calculation is done after measuring the area of all the sections. When experience has been built up, and you have produced a stock of similar designs, it is usual to get the displacement right with minimum correction, simply by comparing the new design at its earliest stages to previous designs.

CENTRE OF BUOYANCY

To find the centre of buoyancy we multiply each section of the Products in the calculation by its distance from the midships section. We simplify by using the number of sections, and not the actual distance, so the second set of multipliers are just 1, 2 and 3.

SECT'N N°	AREA (FT²)	MULTIPLIER	PRODUCT	LEVER	2ND PRODUCT
0	0.00	1	0.00	3	0.00
2	0.32	4	1.28	2	2.56
4	0.87	2	1.74	1	1.74
					Total = 4.30
6	1.49	4	5.96	0	
8	1.35	2	2.70	1	2.70
10	0.90	4	3.60	2	7.20
12	0.00	1	0.00	3	0.00
			Total = 15.28		Total = 9.90

Footnote: These are just typical figures and are found with a planimeter or one of the other methods detailed above.

The second Products are added, and the smaller total taken from the bigger. This gives us
9.90 – 4.30 = 5.60

This is divided by the first total, namely 15.28, to give:
5.60/15.28 = 0.366 feet aft of section 6

The centre of gravity of all the weights must coincide with the centre of buoyancy. To save time working out where all the different weights on a vessel are located, sometimes just the major weights are taken. This short cut works on the basis that the hull, deck and other structural weights are evenly spaced along the length of the craft. (See P40.)

6
WEIGHT & DISPLACEMENT

ARCHIMEDES' LAW & DISPLACEMENT

The weight of a boat and her displacement are the same thing. Archimedes' Law tells us this, and it is a basic feature of naval architecture. This law can be explained by doing a simple experiment:

EXPERIMENT

Take a bowl or tank and fill it to the brim with water. Now gently float a model in it. A quantity of water will overflow. Catch this water as it spills over from the container and weigh it. Now weigh the model – it will be found that the weight of the model is exactly the same as the water collected from the spillage.

From this we can see that the weight of a boat is exactly the same as the weight of water which would occupy the immersed hull space, if the hull was not there.

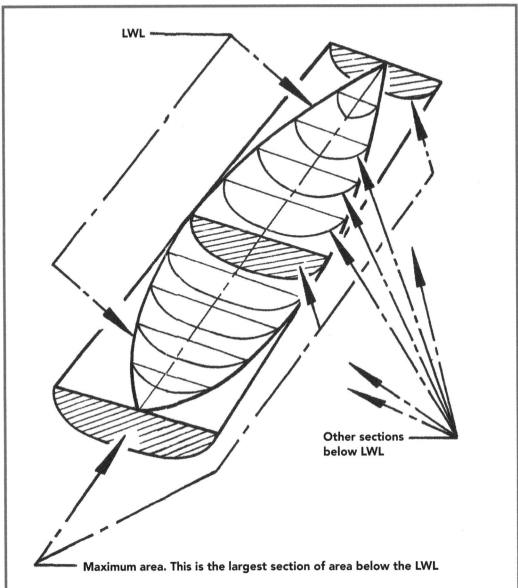

LWL

Other sections below LWL

Maximum area. This is the largest section of area below the LWL

THE PRISMATIC COEFFICIENT
This Coefficient is:
Displacement x 35/LWL x Maximum Sectional Area Below the Waterline (in Imperial Units)
This is a handy coefficient for comparing vessels and as a design guide. Broadly speaking:

SPEED / LENGTH RATIO (Speed in knots / √Waterline Length)	PRISMATIC COEFFICIENT
1	0.525
1.4	0.630
1.8	0.675

ENSURING HULL VOLUME BELOW LWL = TOTAL WEIGHT OF CRAFT

Following on from this, when we draw the lines plan we have to ensure that the volume of the hull drawn below the Load Water Line, measured in cubic feet divided by 35, is the same (in tons) as the total weight of the craft. This is how we know how big to make the hull volume below the waterline.

When a boat is launched she should float exactly on the Designed Water Line. To achieve this, the weight of the yacht has to be known before the lines plan has been completed. (In passing, the weight is taken to include half the contents of the water, fuel and sewage tanks, but no bilge water.)

This determination of the total weight is where skill and experience matter so much. Since the beginner is unlikely to have these talents, it is a good idea to start off designing for fun – when the boats will not be built.

If the next stage is reached, and the beginner's boats are to be built, it is sensible to start with metal craft since these can easily have the final part of the ballast put in after launching. As a result minor mistakes are acceptable in both the total weight and the location of the boat's centre of gravity.

However, even here the final weight must be known to within about 4% when the lines are being finalised. To find out what the weight will be, the new design is compared with other closely similar craft which have been built and float to their marks.

Naturally, similar yachts will have different engines and tank capacities and anchors. These heavy items are taken into account by simply adding together the weights of the heavy items for the new design and the existing boat. If the total of the weights for the new design comes out at say 100 pounds or kilos more than the existing vessel, then the new one must be designed with a displacement that much greater.

In practice the designer may not be able to find yachts closely similar to the one he is planning, so he cannot make a direct comparison. What he does is to draw a graph of roughly comparable boats. They must all be of the same type – all ocean cruisers, or all coastal cruisers, with substantial engines so that they are nearly motor-sailers – or some such criterion.

CUBIC NUMBERS

A 'cubic number' is calculated for each of these craft. Different designers use different 'cubic numbers' according to how their experience has directed them. One way to get this figure is to take the mean of the overall and waterline length and multiply it by the maximum beam amidships and then by the depth of the hull. (Note that the depth of the hull is the distance from the top of the deck on the centreline amidships to the top of the keel immediately below.)

So, where:

LOA = the length overall
LWL = the length on the waterline
B = the maximum beam amidships
D = the inside depth (not the draught} amidships

The formula looks like this:

Cubic number = (LOA + LWL)/2 x B x D

A graph is plotted of cubic numbers against displacements, using at least four larger and four smaller craft to give a credible graph. The cubic number of the new yacht is worked out and her probable displacement is read off the graph.

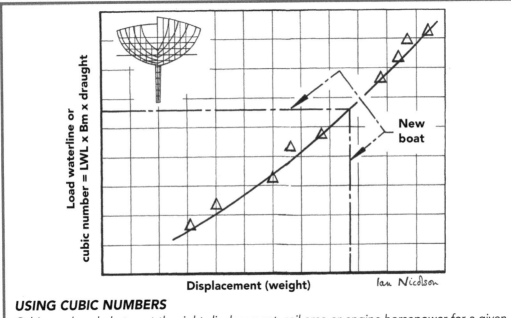

USING CUBIC NUMBERS

Cubic numbers help to get the right displacement, sail area or engine horsepower for a given speed. Here, knowing LOA, LWL, beam, depth (or draught) and weight of 9 comparable yachts, the graph indicates the displacement of the new design.

If the new design has a feature which will put her weight up or down below the average, this is taken into account when deciding on the displacement. Because it is easy to add a little 'trimming' ballast if the yacht floats too high, it is common sense to design the hull with a displacement slightly above the expected weight. Getting weight out of a newly finished boat is difficult and often expensive. There are tricks like drilling holes in the ballast keel and plugging them with wood – but this is tedious and costly.

In any case, items put into the yacht often weigh more than predicted. Furthermore, over the years gear is accumulated so that the total weight after five years is almost always greater than the launched weight. So it's a good idea to have the designed displacement at least 1% and maybe 2% over the expected weight.

CALCULATING A CRAFT'S TOTAL WEIGHT

Up to now finding the weight has been based on using data from existing comparable and successful craft. If such information is not available, or if the boat is a new type, the weight is found by adding together the weights of all the material and components, the engine and all its fittings, the deck gear, mast, rigging – everything.

This sounds a daunting task, but there are techniques which speed up the job. For instance, the weight of a square metre (or square foot) of different thickness of the material used for construction can be checked in a reference book. Then all the areas of the topsides, deck, cabin top and so on which are of the same thickness are listed together. This means that when all the areas have been worked out, the weight for every given area is available.

To work out the weight of the hull, deck and cabin top structure, a rough preliminary lines plan is needed, with the cabin top sketched on. The join line where the thick bottom fibreglass or metal joins the medium thickness, somewhere well clear of the ballast keel, must be drawn in. Then there will be another line where the medium thickness material meets the standard thickness. In this connection wood boats require less work as the planking thickness is the same all over the hull.

To get the total area of the thickest fibreglass or metal, measure the width athwartships round the section. Some people find this difficult but the job can be simplified by marking off intervals of one foot or a third of a metre along the curved lines.

These lengths are added together and divided by the number of sections to give the average length. The resulting figure is half the girth of the thick material. Multiply this by two, as only one side has been measured. Next multiply it by the fore-and-aft length of the thick material, measured round the curve of the buttock lines.

The same is done for the medium-thick material and the thinner material. This whole weight-estimating process makes a designer realise how important it is to keep his plans simple so as to make life easy for the builder. A big array of different thicknesses may be fine in theory, and satisfactory when working on racing craft, but for most vessels there should be a clear tendency to simplify throughout.

When it comes to the equipment, it is easy to omit items accidentally until several weight calculations have been carried out. The trick is to get hold of a major chandler's catalogue or visit their website and work right through it. It is important to remember that the catalogue / website may not include everything which goes on a boat. For instance, most do not list spars, wire rigging or sails. Also a separate engine catalogue / website is needed for the parts connected with the machinery.

It is inevitable that there will be errors in this weight calculation. For instance, some engine makers give the weight of their engines with no cooling water in the passages, and some omit the lubricating oil in the sump. It is therefore sensible to round up figures. Most people work to the nearest kilo or pound, and for craft over 12 metres (40 feet) some work to the nearest 5 kilos or 10 pounds.

What is certain is that a 'high' weight figure is needed, because there is little chance of over-estimating but it is easy to get a figure that is too low. Some designers add an estimated weight for bilge water, others for assorted fastenings, and yet others for the personal gear which accumulates on every vessel ever built.

By adding up all the weights, the total displacement is found. A spreadsheet is used, so that each item is listed with its location. This is usually a distance from the bow or stern. Multiplying the weight by the distance gives the 'lever'. Adding up all the 'levers' and dividing by the total weight gives the fore and aft centre of gravity of the yacht.

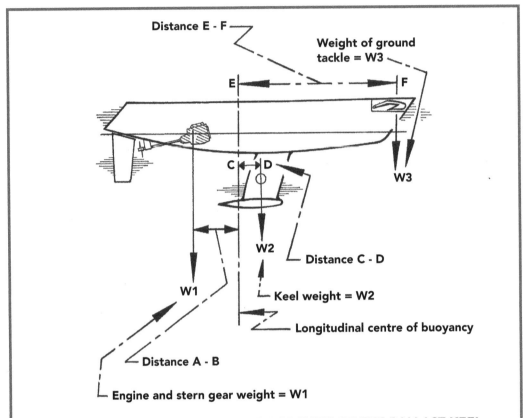

CORRECTLY LOCATING THE CENTRE OF GRAVITY OF THE BALLAST KEEL

This is the traditional way of finding out where the centre of gravity of the ballast keel must be. It is assumed that the centre of gravity of the hull shell, the deck, the furniture, etc., is at the centre of buoyancy. The next assumption is that only the major weights like the anchors and engine will affect the trim of the vessel. Knowing the location of these heavy items it is easy to use the formula:

$$W1 \times AB = W2 \times CD + W3 \times EF$$

The only unknown is CD.
There will be other heavy weights, such as full tanks, the cooker plus the oven, the WC and so on. Their weights and locations are added on the correct side of the equation according to whether they are forward or aft of the Centre of Buoyancy, to expand the formula.

LONGITUDINAL CENTRE OF GRAVITY

It is sometimes necessary to know the vertical centre of gravity, so the vertical height above the bottom of keel of each item is set down and a similar calculation done to the one for the longitudinal centre of gravity.

7

HOW WILL SHE PERFORM?

THE BASIC SPEED FORMULA

A first principle of design is: a moderate or heavy-weight cruiser will have a maximum speed in knots of about 1.4 x the square root of her 'sailing length' in feet.

From this it follows that the bigger the yacht (whether sail or power driven} the faster she will go – other things being equal, which they seldom are.

RULE OF THUMB

Maximum normal hull speed
(in knots)

When not planing = 1.4 x $\sqrt{\text{Sailing Length}}$ (in feet)

This formula tells us that if a yacht has a waterline length of 36 feet, then we take the square root of that number, which is 6 and multiply it by 1.4, giving 8.4 knots. This is known as the 'hull speed'. To attain this speed the vessel must have ample power, either from an engine or from a large enough sail area exposed to sufficient wind strength. In addition, the hull must be the correct shape, and in this connection buttock lines which run steeply upwards aft, at 20 degrees or more, make it more difficult to achieve the speed.

There is another complication: what matters is 'sailing length' and not the static waterline length. So our length figure from which the square root is taken includes some or even all of the length of the bottom of the counter along which the stern wave will curl, and maybe part of the bow overhang. That is partly why our ocean cruiser design has an overhanging counter – to give extra speed in moderate and high winds.

SPEED IN LIGHT WINDS

When sailing slowly, the whole overhang of the counter is above the water. In light airs there is no friction between the water and the hull forward or aft of the Designed Water Line.

If we want a yacht to do exceptionally well in light winds we keep the total area of the immersed part of the hull, keel and rudder as small as possible.

This is known as 'minimising the wetted surface' and the aim is to have the least resistance against forward motion. At low speeds there is no wave-making resistance, just the drag of the water on the wetted part of the hull, keel, rudder and skeg.

A low wetted surface is illustrated on P88 and P89 where we see that the racing yacht hull sections are as near to a semi-circle as possible, while both the keel and the rudder are made as small in area as the designer dares. Because the 'engine' of a sailing yacht is the rig, the sail area is made large if a sparkling light wind performance is the aim.

WINDWARD PERFORMANCE

If a yacht is required to be a wizard to windward, the designer gives her a very deep keel and a tall mast. The former makes it impossible for her to go into lots of harbours but, as in most matters of design, a gain in one area calls for a sacrifice in another.

A high rig needs skillful handling and is expensive, which explains why moderation usually prevails. Lightness aloft is also a bonus, but this puts up the cost. In passing it has to be said that speed costs money which again explains why our ocean cruiser is not excessive in any direction.

Other things being equal, the lighter a boat is, the faster she will be. This explains why our ocean cruiser has a ply deck and cabin top instead of a steel one – weight-saving high up is doubly beneficial because it improves the stability and so gives a double bonus.

Going to windward, two identical yachts will perform equally, unless one is kept more upright than the other. This explains the importance of devices like water-ballast tanks, tilting keels and broad mainsheet horses to give forward drive with the minimum heeling moment. Other requirements are the minimum windage and weight from the deck upwards, as well as the maximum weight as low as possible in the keel.

An ocean cruiser can seldom have a deep keel, so we enhance its efficiency by cramming everything heavy we can down in the fin. She has lead instead of iron ballast because its centre of gravity will be lower and also the space saved can be used for other heavy items like tanks and anchors. Even the engine may be partly down in the fin.

When it comes to the driving forces, the bigger the sail area the more wind is 'captured' so the faster the yacht goes. Under power, other things being equal, the yacht with the most powerful engine will be the fastest until 'hull speed' is achieved. Once this is reached, extra horsepower is a waste of money.

Using standard tables and formulae it is easy to get a reasonably accurate forecast of speed likely to be achieved under power for a given engine – provided the engine is giving the output promised by its manufacturers, the propeller is right, the engine is exactly aligned and so on. It is a good idea to predict the speed with more than one technique and build up data based on actual performance under power.

A heavy boat will not plane, that is 'sledge' along the surface of the water. This means she will never go faster than 1.4 x the square root of the waterline length. Even a vast spinnaker, set in a blast of wind, will not increase the speed as the drag increases as the bow wave moves aft and the hull 'squats'.

PLANING – AT LEAST PARTLY

Up to now, it has been assumed that the speed is limited by the sailing length because this is true for medium weight and heavy craft. However, if the weight can be reduced and the shape of the hull made broad and flattish with a gentle slope at the bow and straight, almost horizontal, buttock lines aft, then, when there is sufficient wind pressure on the sails, the yacht will plane.

She is now no longer just making a trench in the water for herself, but partly slithering over the face of the water.

The concept is easy to explore by taking a flat plank and cutting one end to a sharp bow shape in plan view. Tow it along the edge of a pond and the faster it is pulled the fewer and bigger the waves are which run along the plank sides.

Now fair away the bow by rounding the forward end so that it is sledge-like in elevation. Repeat the towing process; as the speed builds up there comes a time when the bow rises and the plank skeeters along on the surface of the water. This is planing. At this point the rule which states that the top speed achievable is about 1.4 x the square root of the sailing length melts away.

Now the only limitation to the top speed is the driving force which can be applied, the strength of the hull and rig (and the nerves of the crew). The lighter the boat, and the better the planing shape, the faster she will go. However, a planing shape is not ideal for going to windward, so the designer either compromises or fashions a yacht which is a downwind demon, but may be a bit weak to windward.

From a designer's point of view, planing is a mixed blessing because, although it gives extra speed, it means structures have to be stronger to stand up to the increased loads. Strains are proportional to the square of the speed, so the bow area, the hull bottom, the skeg and rudder all need to be beefed up. This increases the weight, so weight has to be skimmed off elsewhere. A vicious circle like this is likely to add to costs.

HOW IS SHE GOING TO PERFORM?

Anyone who has designed a yacht and wants to know how fast she is has to build her first, then sail her well, to find out. However there are short cuts. If a closely similar yacht has less sail area, or less sailing length, or more weight, or a shorter mast, or more wetted surface then the new design should beat her most times.

Once a design is complete, but before the yacht is built, an approach can be made to a test tank like the one run by Southampton University. Here a model of the yacht can be made then towed and tested. The experts at the tank can give predictions as to how the yacht will perform.

It's just a pity this process is not cheap, but then a lot of expertise is involved.

Anyone who designs on a computer and is interested in performance will need a piece of software which gives a 'Polar Diagram'. This predicts speeds on different headings and includes curves for different wind strengths. When comparing different yachts, one of the first things any new designer must realise is that few boats are brilliant all-rounders. This means that the probable weather and wind direction for a given race or series or cruise has to be taken into consideration before, during and after doing a design.

8

SAIL PLAN

CENTRE OF LATERAL RESISTANCE

Before drawing the sail plan the designer finds out where the 'Centre of Lateral Resistance' is located. This concept, called the CLR for short, is the 'centre of gravity' of the yacht's profile below the waterline. The whole matter can be visualised by thinking about a simple experiment.

EXPERIMENT

Suppose the yacht is to be pushed bodily sideways through the water by a boathook. If the boathook is too near the stem, the yacht will pivot and the bow will be pushed away. If the boathook is then moved too far aft, the stern will go off and the boat will turn with her bow pointing towards the person shoving the boathook. However if the boathook is jammed against the hull at precisely the right spot, the force on the boathook will shove the boat bodily sideways.

The boathook is then at the CLR.

The CLR can be found by mathematical methods or by computer programmes made for finding a centroid, but a lot of people prefer the old-fashioned technique because it is quick, reliable and almost impossible to get wrong.

It works like this:

- A tracing is made of the elevation below the waterline including the skeg, but not the rudder or stern gear such as the propeller and its shaft
- This shape is cut out (the tracing paper should be of a heavy type)
- The paper is folded parallel to the waterline several times till it forms a rigid 'rod'
- This 'rod' is laid on a knife blade till it balances like a see-saw, and a mark is made there
- This is the CLR and its distance from the fore end of the waterline is marked on the lines plan

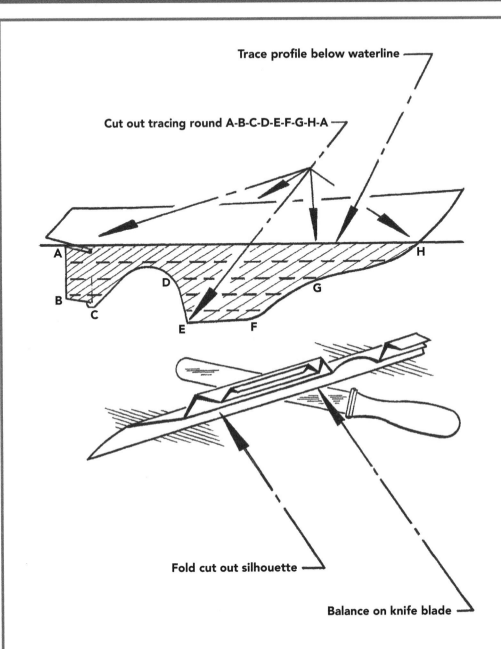

Trace profile below waterline

Cut out tracing round A-B-C-D-E-F-G-H-A

Fold cut out silhouette

Balance on knife blade

FINDING THE CENTRE OF LATERAL RESISTANCE

A simple way to find the fore and aft position of this point is shown here. Make a tracing of the underwater part of the vessel which is the shaded part of the sketch above. Cut around it and then fold along dotted lines, fore and aft, as shown.

The bottom part of the illustration shows how the folded profile is rested on a knife blade so that it balances. Mark this point. This is the Centre of Lateral Resistance – the CLR. To transfer the point to the sail plan, measure the point from the fore end of the waterline and transfer the distance to the sail plan.

CALCULATING THE SAIL AREA REQUIRED

The designer decides what total sail area is needed by looking at comparable craft, or by using the sail area graph in *The Boat Data Book*. In this instance the 'area' is taken as the combined areas of the mainsail and the 'fore triangle'.

The latter is the distance from the fore side of the mast at the bottom to the bottom of the outer forestay multiplied by the distance from the deck to the top of the outer forestay, all divided by two.

Another way to decide on the required sail area is to use the formula:

Sail Area =
DWL x Maximum beam at waterline x C

The factor C varies between 1.8 to 3.2 for a wide range of yachts from dinghies up to 18 metres (60 footers). Fast racing boats have a high factor, motor-sailers intended for retired people have a small C. A heavy boat needs a big sail area but for long range cruising a snug rig is needed for safety. Big sail areas cost a lot, but small ones mean poor performance in light winds.

Most yachts have a Bermuda sloop rig because it is efficient, well developed, simple and the components are widely available at keen prices. Specifying a different rig, such as a ketch, can be justified if the sails of a sloop are too big for the crew to handle.

On cruisers below 12 metres (40 feet) it is often a requirement that a single person can handle every sail on board:

- A fit experienced man can handle a sail up to about 46m² (500ft²)
- Someone who leads a sedentary life will find a sail about 35m² (375ft²) as big as can be managed in severe conditions
- Children and creaky pensioners know a 23m² (250ft²) sail is their limit in rough weather

In theory, modern handling equipment such as roller furling sails, full length battens, as well as multi-speed and power-operated winches, have increased these limitations. But no one has told the sea it's time it was tamer and in severe conditions things go wrong. So these sail sizes are only exceeded if adequate precautions are taken.

The mast is kept as short as practical to enhance stability and keep down its price. The boom must be short enough to clear the backstay even when it lifts above horizontal during a gybe. As bowsprits are expensive and vulnerable, it is normal to have the forestay fixed at the stem head. These limitations dictate how the sail plan is drawn.

RULE OF THUMB

Sail Handling Limits

A fit experienced man	46m² (500ft²)
A more sedentary person	35m² (375ft²)
Children & the more elderly	23m² (250ft²)

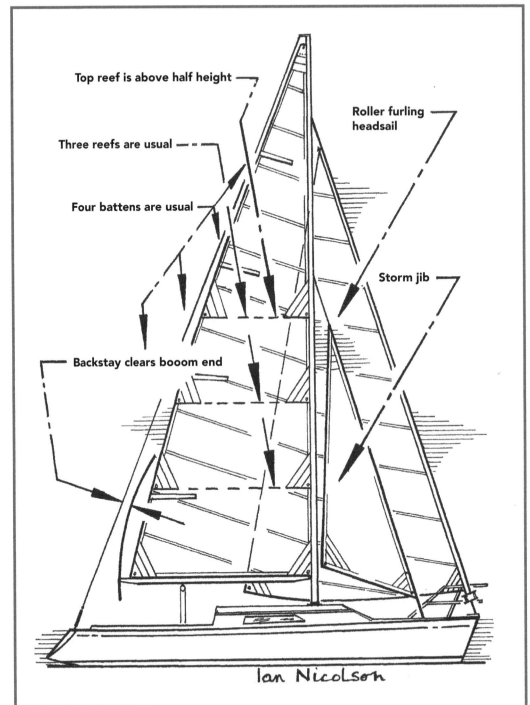

Top reef is above half height

Three reefs are usual

Four battens are usual

Backstay clears booom end

Roller furling headsail

Storm jib

Ian Nicolson

SAIL PLAN DETAILS

A typical sailing yacht has these features. Three reefs are found on all but the smallest craft, the top one being designed to deal with force 9 winds.

Even when an unexpected gybe lifts the aft end of the boom it should never clip the backstay. A separate stay for setting the storm jib is usually needed.

DRAWING THE SAIL PLAN

To begin the sail plan, the hull profile above the water is taken from the lines plan and the CLR marked on the waterline. The scale used will often be half of that of the lines plan. The outlines of the cabin top, main hatch hood, vents and other major deck gear are added.

The lines of the cockpit sole and seats are drawn. On yachts above 11 metres (36 feet) the crew want to be able to stand in the cockpit and not be hit by the boom, so a horizontal line 2 metres (6 feet 7 inches) is drawn above the cockpit sole to show the bottom of the boom. Few people are as tall as this, but below the boom there are bails for the mainsheet and such-like which must clear the crew.

On smaller yachts the crew must be able to sit in the cockpit without being touched by the boom, so a horizontal line 1.09 metres (3 feet 7 inches) is drawn above the level of the seats to show the bottom of the boom. The top and bottom of the boom are drawn in, taking the boom diameter from spar tables.

> **RULE OF THUMB**
> **Minimum height of the bottom of the boom**
>
> Yachts > 11m (36ft): 2m (6ft 7in) above the cockpit sole
>
> Yachts < 11m (36ft): 1.09m (3ft 7in) above the level of cockpit seats

The aft side of the mast at deck level is drawn initially at 38% of the waterline length from its fore end. The back of the mast is now drawn to a height above the waterline equal to 1.35 times the overall length of the boat, for a typical fast cruiser. For a low-rigged offshore cruiser this figure is brought down to 1.15 times the overall length. At this stage the mast is raked aft 2 degrees. The mast's forward edge is drawn in, taking the spar diameter from the table of suitable tube dimensions.

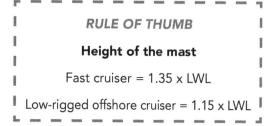

> **RULE OF THUMB**
> **Height of the mast**
>
> Fast cruiser = 1.35 x LWL
>
> Low-rigged offshore cruiser = 1.15 x LWL

The outer forestay is drawn in from the top of the stem to the masthead – or to a lower point on the mast for a 'three-quarter' rig.

The mainsail leech is drawn starting from a point on the aft edge of the mast 2% of the mast length down from the top. This distance down allows for the halyard sheaves and the topmast fittings because these prevent the sail from extending to the very top of the spar.

The clew of the mainsail needs some form of outhaul at the aft end of the boom, and this needs space. As a result the mainsail clew has to be located about 2% of the booms length from the aft end of the boom. This establishes the bottom of the leech.

The basis of the sail plan is now in place and it should be studied to make sure it looks right and is what the designer is seeking. Does it look correct compared to similar craft produced by competent naval architects? Are the crew going to be able to manage the sails? If not, the mainsail will have to be reduced in area and a staysail fitted as well as a jib. The mast will probably need moving aft for this, and the boom may need shortening to avoid fouling the backstay.

CALCULATING THE CENTRE OF EFFORT

Next the 'Centre of Effort' (CE) is calculated, as shown in the caption below. This is the centroid of the fore triangle area and the mainsail area combined. If there is a mizzen, this is added too. The CE is marked on the sail plan and the horizontal distance between it and the 'Centre of Lateral Resistance' is measured.

This distance should be about 16% to 18% of the waterline length, but wide boats need more, sometimes up to 20%.

Short of going to the expense of tank testing, a beginner has problems deciding what percentage of the waterline length the CE should be forward of the CLR.

There are two ways forward: copy similar boats by acknowledged masters, or incorporate arrangements for shifting the mast fore and aft. Typically this movement will be of the order of 2% of the boat length. The cost of construction will be increased, but that is far better than ending up with a yacht which is heavy on the helm.

If the mast position is variable, the chainplates must be of the horizontal bar type on deck, with a selection of holes for the shroud ends, so that, as the mast is moved, the shrouds can follow.

At this stage the designer has to complete the sail plan. The mast was located 38% of the waterline length from the bow just as a preliminary guide. If the 'lead' is too small the whole rig may need moving forward, or the foot of the mainsail may be shortened, or the forestay bottom end can be moved forward by fitting a bowsprit. Adjusting the mast rake will make little difference, and it is better to leave it at 2 degrees, as then it can be varied during trials by way of fine tuning.

At this stage a beginner is advised to draw out several sail plans incorporating differences. They should be pinned up on the wall and studied over several weeks.

CENTRE OF EFFORT (CE)

The CE of the sail plan is found by first getting the CE for each sail. The luff of the mainsail is divided in two, each part being D in length. The middle of the luff is joined to the opposite corner, the clew at the aft end of the mainsail. Next the mid point of the foot of the sail is found, and this is joined by a line to the head of the sail. This line cuts the line from the luff to the clew at K. This is the Centre of Effort of the mainsail.

Follow the same proceedure for the headsail, the result is at J. Join K and J by a line, which is divided so that KZ is a length proportional to the headsail area and JZ is proportional in length to the area of the mainsail.

An easy way to divide up KJ correctly is to draw vertical lines at K and at J. Now measure off a length down from K proportional to the area of the headsail to give KN.

Next measure up from J a distance proportional to the area of the mainsail, and mark the point M so that JM is proportional to the area of the mainsail. Join M and N, this meets KJ at Z which is the Centre of Effort of the whole sail plan.

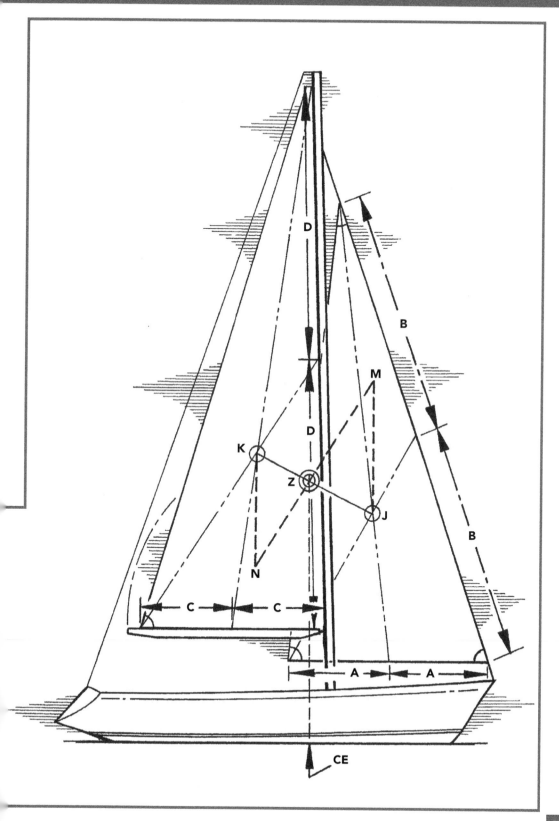

9
CONSTRUCTION

As with other chapters in this book, this one needs amplifying by reading the books listed in the Bibliography, and articles in the technical press. Boat construction is a vast subject, and it often takes far more time to do a full set of construction drawings than any other plan.

CONSTRUCTION DRAWINGS & DETAILING

The first move is to trace from the lines plan the outline of the yacht in plan and elevation, plus sections at critical locations. Here, as with other drawings, it is a good idea to trace the outline in ink so that later, when erasing mistakes with a rubber, the outline is unaffected.

There are designers who only do a plan and elevation of the construction, but the best do plenty of sections too, and often a selection of small drawings to explain details. These subsidiary drawings are done at a scale of 1/4, or better still, full size. They fill the spaces around the main drawing, or cover separate sheets of paper, or both.

The designer who wants to be a success with boatbuilders takes special trouble with the construction plans, and does not stint on the extra sketches to clarify details. The specification, which is the written description of the building, is important, but plans matter more because they tell a fuller story.

For each of the materials that are used to build boats there are many different construction techniques. Wise designers visit the boatbuilder first to discuss his favoured methods. These should then be incorporated into the construction plans, preferably with some new enhancing details.

SOME ESSENTIAL PRINCIPLES OF CONSTRUCTION

A beginner should avoid radical changes to well-established building processes because they have been refined by time and experience – unless, of course, the beginner is a genius.

Craft up to about 7 metres (23 feet) may be built with no internal stiffening and this is called 'monocoque' construction.

Here the hull-deck unit is like an eggshell in that the external structure, which keeps the water out, has sufficient strength to be self-supporting and withstand expected but not severe accidents. Steel craft up to 9 metres (30 feet) or even more are built by this method, especially for inland waters.

The more usual approach is to have internal strength members which support a relatively light 'skin'. The stiffening may be athwartships frames, or fore-and-aft stringers, or a combination of these. Bulkheads are usually incorporated as strength members, and it is common to use some of the furniture to help keep the hull rigid and be capable of withstanding buffeting.

A lot of yachts are built by a combination of 'monocoque' and framed construction, in that the areas which are not likely to be heavily stressed have nothing but the hull shell to give them strength. Where the loads are high or accidents likely, there are stringers, frames, and even perhaps diagonal stiffeners. Sometimes the furniture is made extra strong and well-secured in place.

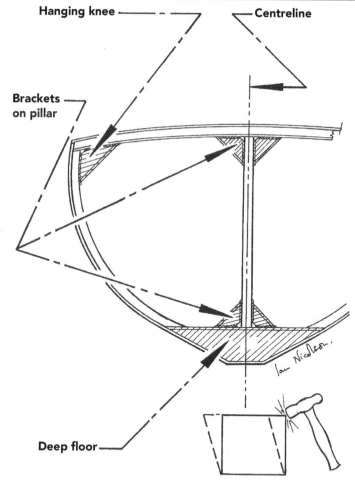

Hanging knee ——— - ——— ——— Centreline

Brackets on pillar

Deep floor ———

STRENGTH IN ENGINEERING

Boats need structure to prevent them distorting when they hit a wave or quay wall. The small sketch above shows what happens when a box is bashed at one corner.

The mid-section of a strong aluminium or steel yacht is shown with brackets and hanging knees as well as a deep floor to give massive strength.

FIBREGLASS CONSTRUCTION

Anyone new to designing is not likely to be involved in the usual method of building a fibreglass vessel, because it is geared towards mass production. This is because boats are built in moulds and the costs of producing the moulds are considerable. However, a beginner wanting to build a single new boat to his own design may go for foam sandwich fibreglass construction because this method of building suits one-offs. As a guide a typical semi-mass midship construction section is shown below, and as always with boatbuilding there are a great number of variations on this theme.

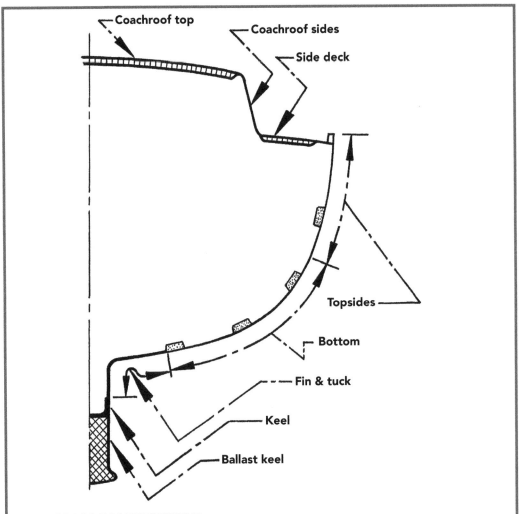

FIBREGLASS CONSTRUCTION

This typical construction mid-section is of the Bowman 48. The coachroof is of 3,150gsm (grams per square metre) fibreglass plus 19mm (3/4 inch) balsa core. The coachroof sides are 3,750gsm and the side decks are the same as the coachroof except that the balsa is 12mm (1/2 inch) thick. The topsides are 3,450gsm and the bottom 5,250gsm. At the fin and tuck the thickness is 7,050gsm but down at the keel it is 8,850gsm.

The plans must show what materials are to be used, and where. There will have to be extra stiffening down towards the keel and, as with all forms of construction, there must be no sudden change of thickness or strength and ideally no sharp 'change of direction'.

Where there has to be an angle change, for instance at the transom edge, or the join in the more or less horizontal deck to the topsides (which are virtually vertical) there must be extra structure to handle the loss of strength. The approach usually taken is to thicken the skin and use bulkheads to take some of the stresses.

For economy and to save time, it is usual to use the same thickness of foam all over the hull and deck, but increase the fibreglass thickness (inside and outside):

- Below the waterline
- Halfway down the curve from the waterline to the keel
- By the keel

The designer needs to show all these boundaries.

OCTOPUSES, GRIDS & WAFFLES

To cope with the severe strains imposed by the ballast keel when the yacht is thrashing to windward there is a strong structure with fore-and-aft as well as athwartships components incorporated inside the bottom of the hull. It is called variously an 'octopus', 'grid' or 'waffle', and it may be made up of floors (which are like deep frames under the cabin sole) combined with settee fronts.

Alternatively there may be a separate moulding of fibreglass bonded to the hull. Usually the 'grid' incorporates the structure which supports the bottom of the mast, mast pillar or mast support bulkhead.

BULKHEADS

Bulkheads of marine ply are commonly used to strengthen the hull and take loads from such things as the shrouds, inner forestay, engine bearers and so on. The bulkheads are often made of 12mm (1/2 inch) ply regardless of the type and size of cruiser as it needs little if any strengthening to take the chainplates and so on. It's not too pricey. It's widely available and boatyards tend to build up a stock of this material so that there are plenty of off-cuts and odd ends available.

Doorways in bulkheads are cut with rounded tops and bottoms to minimise the loss of strength. After the hull shell is complete the bulkheads are glassed in all around the edge. Among the copious details on the construction plan, the designer shows the extent and type of this glassing.

WOOD CONSTRUCTION

The old-fashioned way of building wood boats takes a lot of time, skill and money. Variations have been invented which aren't quite as costly, such as strip planking. Here the shell is made up of longitudinal parallel-sided strips which are glued together along their edges, sometimes with nails driven down through one plank into the next; a layer of fibreglass on the outside for extra strength and a lasting attractive finish may be added. Bulkheads and deep frames, typically about 1 metre (39 inches) apart, form almost all the internal framework apart from the keel support 'grid' and mast support components.

A backbone of wood has to be built first and it will normally be made up of laminations, which are fairly thin planks glued together with the joins 'staggered'. This 'staggering' is a basic principle of all construction and involves separating adjacent joins by the biggest practical distance. Furthermore joins should not be located at bends.

The backbone runs as a continuous structure from stem head to transom, with curved joins. Where curves are not possible, such as at the meeting with the transom, brackets or knees are used instead, as with other forms of construction.

STEEL CONSTRUCTION

Steel construction is simple, fairly cheap, easy to repair and ideal for 'one-off' craft of all sorts. It is a good material for a new designer because the problems have, for the most part, been solved long ago. Before designing in this material it makes sense to read *Small Steel Craft* (see the Bibliography).

The construction plan starts with the backbone which may be of plate or extruded bar, rod or tube. Here, as with all design jobs, clever draughtsmen first contact several firms selling the various building materials to establish what is available, economic and of the right quality. Clever designers aim to use no more than three different plate thicknesses and three types of extruded bar.

Bright operators visit steel suppliers and go through their stock lists. There are locations like the bottom of the fin and the main keel plate where relatively small quantities of thick plating are needed, and buying off-cuts can save a lot of time waiting for deliveries, and lots of cash too.

The top of the stem bar may change from cheap mild steel to stainless steel because the forestay is fitted here and there will be constant wear at the rigging screw toggle which would chafe through paint and soon produce rust. For the same reason the chainplates, backstay plate, fairleads, bollards and other steel parts on deck are made of a stainless material.

For quick, and therefore economic, building there are a small number of strong frames which are either bent from flat bar or cut from plating. They have a wide athwartships dimension to allow close-spaced flat bar stringers to be slotted into them.

Many designers prefer to use T-bar or angle-bar stringers because they are stronger for a given weight, but they are very awkward to shot-blast and paint on the outboard side, so eventually they rust. These stringers are partly to stiffen the plating, and partly to help the builder when he is wrapping the plating on as he forms the hull.

A good way to show a flat bar on edge on drawings is to use a thick pencil line. Some draughtsmen show a single line for angle-bars but this can be confusing for the builder, even if he appreciates that the line is drawn at the junction of the flanges. Besides, angle-bars take up space and if their outside limits are not drawn another item may be put in where, in practice, there is no space.

A designer has to make hundreds of decisions for every line he draws. A typical dilemma is the construction of the deck. One line of argument recommends a steel deck because it results in a submarine-like boat, all of steel, with lots of strength and, in theory at least, no chance of leaks.

However a ply deck bolted all round the edge saves weight and the need to have under-deck lining, makes it easy to add or alter deck fittings, acts as a good insulator, is quickly fitted, the underside can have all but the last coat of paint applied before the ply is laid down and so on. For comparable reasons, wood beams are sometimes fitted except in way of the mast, at the fore end of the cockpit and the aft end of the foredeck where extra strength is essential.

From this it will be seen that, when doing any construction plan, there are few rules – apart from the need to use good engineering practice. For instance, all structures should be triangulated and all strong parts tapered away at the ends. It is also essential to remember to use large printing because plans in boatyards get rough treatment, so anything indistinct soon becomes illegible.

Experienced designers insist that all drawings sent to boatbuilders are immediately fixed onto large ply boards so that no one can roll them up, or stuff them away in a drawer and forget where they are!

ALUMINIUM CONSTRUCTION

Broadly speaking this is comparable to steel. As a rough rule, boatbuilding aluminium alloys are one third the weight of steel and two-thirds the strength. Only their cost prevents them from being more widely used, and some owners save money by leaving the material unpainted, especially inside the hull.

The designer produces drawings which are comparable to those for a steel vessel but precautions are needed to cope with vibration as this results in weld fractures. Over the propeller, for instance, the plating has to be thicker than elsewhere and have lots of stiffeners. The P-bracket needs much reinforcing and the engine bearers are bracketed extensively. The designer specifies, here as elsewhere, that the welds must go right around the ends of every bracket.

Aluminium is fairly soft. This makes it vulnerable to wear, so by fairleads there have to be half-round-section chafe strips welded on, or some similar precaution taken. Chainplate holes must be bushed with stainless steel and precautions taken against corrosion.

Electrolytic action, which is corrosion caused by seawater and another metal, is the big problem and, before beginning the construction drawing, read Nigel Warren's book *Metal Corrosion in Boats*.

SUMMING UP

More mistakes are made in the construction drawings than on the other plans. Many of these mistakes are sins of omission because, if something is not shown in detail, the boatbuilder (who has a lot of trouble making a profit) will follow the easiest route.

Regardless of which material is being used, the designer has to be up-to-date with the latest techniques. Then he has to avoid using them, unless he is building a racing boat! The newest building method is one that has not been fully developed and whose snags have not been ironed out.

A principle skill in designing is to know how far to go in modernising a construction plan. Anything too complex or too innovative and the builder will either quote a massive price for putting the yacht together, or he will go bust building her. Anything too old-fashioned and the work will be too laborious... with the same result.

10
ENGINES

It is usual to fit an inboard marine diesel in a yacht, unless she is small, in which case an outboard is often used. In this instance 'small' is an elastic term, and may be anything below 9 metres (30 feet) for a light cruiser-racer, and anything below about 7.5 metres (25 feet) for a genuine cruiser.

Modern yachts' diesel engines are reliable, economical and stand up to so much abuse that they will often last 20 years provided they are given basic maintenance. If they are well-loved and have freshwater cooling then an even longer life can be expected.

FACTORS INVOLVED IN YOUR DECISIONS

There are many ways of selecting the right engine for a boat, ranging from simple rules-of-thumb to complex formulae and easy-to-read graphs. For instance some designers go for 1 horsepower (hp) per 2.3m² (25 square foot) of sail – and so they fit an engine of about 12hp in a yacht with 27.9m² (300 square foot) of sail. In this instance the area is taken as the mainsail and the fore triangle.

This formula is an over-simplification. Irritatingly, it puts a big engine in a cruiser biased towards racing which tends to have a large sail area, and a smaller engine in a heavy long-range cruiser with a safe small sail plan. This is the wrong way round, because cruiser-racers are generally lightly engined whereas deep-sea craft are better with ample spare horsepower to enable them to battle against strong winds and rough seas.

In passing, this whole matter of powering a yacht has a lot of complications, some of which are ignored here, but are discussed in detail in the books listed in the Bibliography. For instance, when designing an ocean cruiser it is good practice to select an engine which is used worldwide so that you can count on the fact that spares and knowledgeable mechanics are widely available.

Anyone new to naval architecture can save a lot of time when practising and doing preliminary design work by using the

RULE OF THUMB

Choice of Engine:

When selecting an engine for an ocean cruiser, choose a type which is used worldwide so that spares and servicing are available as and when needed.

weight of the yacht as the basis for selecting the right horsepower. There is logic in this since the engine is not required to make a sailing yacht plane, so the maximum power needed is that which will give 'hull speed' in knots, namely 1.4 x the square root of the waterline length in feet (see P41).

Generally speaking an engine which has 3 horsepower per ton of displacement will get near hull speed, and 4hp per ton should be ample in calm water. In practice, lots of designers go for 5hp per ton for cruisers, and some select 6hp per ton to cope with things driven off the engine. These can include an extra alternator for quicker charging of the batteries and in case the main alternator breaks down. There may also be a power bilge pump which doubles as a deck wash pump and fire pump.

If a hydraulic anchor windlass is fitted the driving motor will normally be powered by the main engine. Freezers and fridges need power and so do air compressors – the list grows with each generation.

Broadly speaking, horsepower is cheap, in that there is not a vast price difference between a 50 and a 70hp engine. A bigger propeller will be more expensive, but the cost and size of the exhaust, silencer, starter battery and controls is unlikely to increase much if at all.

It may be necessary to have a propeller shaft that is one size larger which will entail using a bigger stern tube and gland. The price increase here will be noticeable, but it will be small when taken as a percentage of the total cost of the installation. Broadly speaking it does not cost a great deal more to increase the engine output by a useful amount, especially as the charges for the installation work will be almost identical.

This extra punch is seldom used, then one day the yacht is badly caught out. She may be on a lee shore, or in a crowded harbour. It is seriously windy and everyone is worrying – then next moment there is a hurricane blast. The only craft which can cope with this, and make progress in spite of steep high waves and the gigantic thrust of the wind on the spars and rigging are those with lots of spare horses. These craft sometimes have 8 or more horsepower per ton of displacement.

Most of the time these big engines are lightly loaded. Their propellers usually turn fairly slowly – however there is a tiger lurking in the bilge for the day when there is a life-or-death fight, and that is comforting.

An advantage of an extra big engine is that it is often only run gently, so it is not too noisy. Broadly speaking marine engines running below 1,500rpm will not spoil a night's sleep – the disturbance factor rises with the engine revs. Also the big beefy engine can have extras added on in later life without affecting the yacht's performance.

It is not so long ago that designers were putting in engines giving 2hp per ton, and that was fine if the only requirement was to get home in a flat calm. Some craft were fitted with tiny engines, with just 1hp per ton, mainly to recharge the batteries and to give steerage way out of harbour. These little power plants take up no room but they are too small for most owners – even dedicated racing enthusiasts.

PROPELLERS

Once the engine has been chosen, the correct diameter of propeller is selected from the relevant graphs. If these are not available, the engine maker or his local agent will tell the designer the size of propeller needed, based on the size, type and weight of the yacht.

Between the tip of the propeller blade and surrounding structure there must be a distance equal to 12% of the propeller diameter to prevent excess noise and possibly hull damage. During the early stages of a design it is unwise to draw the centre of the propeller with just the minimum clearance.

So the procedure is to write down the diameter of the propeller and halve it. Now add 16% or even 22% of the diameter –

to be safe and also to allow for a future change of engine when a larger propeller may be needed. This addition gives us the distance down from the underside of the hull shell to the centre of the propeller.

> ### RULE OF THUMB
>
> **Minimum Clearance for Propellers:**
> There must be a minimum clearance of 12% of the prop's diameter from blade tip to any surrounding structure.
>
> **Better Practice:**
> Use a clearance of 16-22% of the prop's diameter to allow for the time when a larger prop is needed.

PROP SHAFTS & ENGINE POSITIONING

This centre is also the centre of the propeller shaft. Draw a horizontal line from this centre, forward right into the hull. Draw a second line starting from this centre and angled up 8 degrees above the horizontal – again extending forward into the hull.

The majority of marine engines should not be run at an angle steeper than 10 degrees, otherwise the oil pump cannot reliably suck oil from the sump. However, when a yacht is driven hard she tends to squat by the stern, especially with the crew aft in the cockpit. So, to be safe, do not work up against the 10 degree limit but keep two degrees, and ideally more, in hand. Hence the 8 degrees limit line.

The elevation of the engine is drawn on a piece of tracing paper to the same scale as the yacht. Some engine makers supply these silhouettes on tracing paper to different scales. The centreline of the shaft is marked on this tracing and it is aligned with the horizontal line from the centre of the propeller.

If the engine can be installed with the shaft horizontal the installation is easier, and therefore quicker and cheaper. This ensures that there should never be a risk of oil starvation even if the yacht is badly out of trim. In some long range cruisers fitting the engine amidships and partly down in the keel means that there is no

> ### RULE OF THUMB
>
> **Angle of Engine Mounting from the Horizontal:**
> Most marine engines should not be mounted at an angle any steeper than 10° from the horizontal otherwise the oil in the sump will not be circulated.

trouble arranging a horizontal shaft. Also the engine weight is extra low – another bonus.

Usually the engine is located under the cockpit, so it is tilted down at the aft end. Generally designers want the engine as far aft as possible, to maximise space in the cabin. However there must be room inside the hull, aft of the engine, for the stern tube, the stern gland and the engine coupling.

It is well worth drawing on the engine silhouette the coupling, then allowing a minimum distance (to scale) of 100mm (4 inches) of propeller shaft, followed by a standard stern gland and its flexible coupling, plus a length of at least 150mm (6 inches) of stern tube inside the hull. This establishes the furthest aft the engine can be fitted inside the hull.

The bottom of the engine and gearbox must all be 50mm (2 inches) clear above the inside of the structure, not just above the outside of the hull. Tracings of section views at the front and back of the engine are laid on the appropriate construction sections of the hull to confirm there is clear space everywhere under the engine.

It is not safe to take the section off the lines and draw in the thickness of the structure at the engine's intended location. Boatbuilders sometimes ignore plans, and when installing the engine, as with other stages in the construction, the designer has to assume that at least a small error may be made.

Though the construction plan says the hull thickness in way of the engine is 15mm (5/8 inch) of fibreglass, someone may have been heavy-handed so it may be 18mm (3/4 inch) thick – with upstanding globules of fibreglass an extra 6 or 12mm (1/4 or 1/2 inch) high.

There is more. When fitting the stern tube the builder may get the angle slightly wrong. So experienced designers tend to allow more than the minimum 50mm (2 inches) clearance between the hull and underside of the machinery. This extra space also makes it possible to clean under the engine, salvage dropped tools and perhaps drain the oil via the sump drain plug.

Engine bearers are drawn in so as to be as long as possible. Some factory-built yachts have bearers which are too short, so that there is more vibration than necessary and sometimes cracks appear at the ends of the bearers. Each end of the bearers should be tapered out, regardless of what material is used.

To form a sump under the engine and retain any oil that has leaked, it is usual to join the ends of the bearers with athwartships floors. Wood yachts need a metal drip tray under the engine to catch oil leaks.

The casing is drawn around the engine allowing room for the sound-proofing. Space is needed for air circulation and access. If the engine is to be fitted in the saloon the whole casing should be portable and in easily-handled sections.

THE EXHAUST, SILENCER & CONTROLS

Remembering that there are few bad engines, but plenty of bad installations, the designer sets about reducing the chances of crises.

He draws the line of the exhaust so that it goes downhill from the engine to the silencer at about 45 degrees. The cooling water enters the exhaust at the aft end of the engine manifold. There must be an adjacent structure to take the silencer's holding down bolts and clips to hold the exhaust piping.

Out of the aft end of the silencer the exhaust sweeps up to close under the deckhead or, better still, up inside a cavernous cockpit coaming so it is as high above sea level as possible. From there it extends steeply down and athwartships to prevent waves washing into the silencer when the engine is not in use.

The outlet is aft near the waterline but not so low that it is covered by the stern wave or average waves.

An exhaust pipe must have no sharp bends because they cause back pressure and loss of power. Rubber exhaust hose can only be bent round gentle curves which typically have a radius about 16 times the diameter of the piping.

The gear and throttle controls must be located so that all the components are accessible for lubrication, but where the gear lever will not be obtrusive. It is best to have the engine panel with the rev counter, starter key, oil pressure gauge, etc. in a cockpit locker or by an inside steering position because this prevents weathering and corrosion. However there must be an audible alarm to deal with overheating because when control panels are not under the helmsman's eye they cannot be monitored constantly.

DECISIONS TO BE MADE 'AT SHIP'

Some aspects of the engine plan are not finalised until the engine is being installed. It is better to make some decisions 'at ship' rather than on the drawing board so that alternatives can be examined.

This applies to other parts such as deck fittings, but it is poor practice to write too often on the plans 'Fit at ship' because it holds up the building and may add to costs.

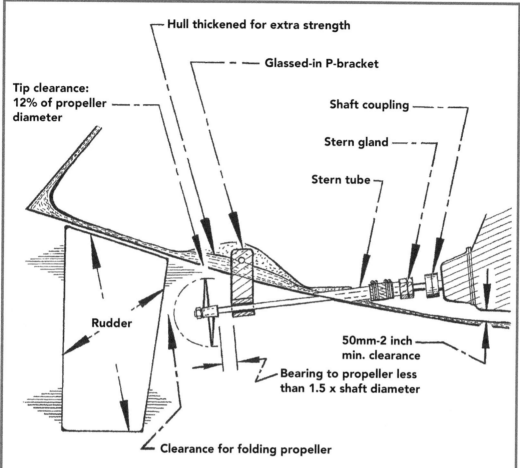

Hull thickened for extra strength

Glassed-in P-bracket

Tip clearance: 12% of propeller diameter

Shaft coupling

Stern gland

Stern tube

Rudder

50mm-2 inch min. clearance

Bearing to propeller less than 1.5 x shaft diameter

Clearance for folding propeller

FITTING IN THE STERN GEAR

It is important that there is clearance around the propeller, space for the stern gland and coupling, as well as a minimum space under the engine. This drawing shows that, regardless of the size of the vessel there must be 50mm (2 inches) under the engine to allow for irregularities in the construction and a flow of cooling air.

The propeller tip clearance shown is usually considered the minimum and the distance between the bearing and propeller a maximum.

11
THE CABIN OR ACCOMMODATION PLAN

USING ONE METRE INTERVALS

When doing the preliminary of the cabin layout plan it is often a help to divide the length of the vessel up into 1 metre long (3 feet 3 inches) sections. This speeds up the allocation of space because settees and berths need to be 2 metres long.

In the same way, a small galley or a comfortable toilet space occupies about 1 metre length fore and aft. The length also suits a chart table space, broadly speaking. At this initial stage the designer is working out the location and juxtapositions of the compartments and furniture, so this quick '1 metre space allocation' is a handy design tool.

For small boats of, say, between 6 and 11 metres, a cockpit or wheelhouse of 2 metres fore and aft is often about the right length. For yachts of the next size up, the cockpit will often be satisfactory if it is roughly 3 metres long.

A chain locker of 1 metre fore and aft is ample for most craft especially in the very narrow bow space where there is room for little else.

A navigating bridge or steering position 2 metres long suits many craft, except very small ones where a 1 metre length will be found adequate if space on board is very short.

Naturally this technique needs to be approached intelligently and must not be used rigidly. But it speeds up the preliminary planning stage and is easily adjusted for the final design. For instance, the forward cabin may have to have 'king-sized' berths and for this it will usually be possible to take a little length from the chain locker at the fore end, or from the toilet space just aft, or from both.

DESIGN WITH THE CREW'S SAFETY IN MIND

When arranging the below-deck spaces it is important to remember that, at sea, small craft are thrown about a lot. Consequently the cabin sole should not be so wide that the crew cannot cross it safely in rough conditions. Likewise there must be plenty of grabrails. Furniture must not have any sharp edges for safety reasons.

KEEP IT SIMPLE

There will often be a temptation to have a novel or unusual arrangement or layout of the cabin space. This may be risky because the established practices have been worked out over many years and have been refined by experience.

There is no reason why a 'normal' or common layout should not be used; it can always be enhanced by the use of special decorations or the use of special materials and finishes, or by having some features which are styled, such as laminated components, carving or buttoned upholstery.

IMPLICATIONS OF BOAT MOTION FOR THE LAYOUT

When planning the internal layout it is important to remember that the motion on board will be least about two-thirds of the length from the bow and worst right forward.

The toilet is seldom located at the very bow and the galley should be placed where the crew will have the best working conditions.

At sea, sleeping right forward can be impossible even for hardened crew, so sea-going berths are normally at least aft of midships, and sometimes right aft. They should not be near a noisy engine.

BULKHEADS

It is normal practice to reduce accommodation costs by having bulkheads extend athwartships across the vessel at right angles to the centreline. They should be neither stepped nor angled. This practice assists the strength of the yacht as a whole as well as reducing costs. All doorways through bulkheads should have rounded corners, again for reasons of structural integrity.

STANDARDISE COMPONENTS

Components should be standardised as far as possible. For instance, drawers should all be made to the same size throughout the vessel, or at least restricted to two sizes. The same applies to locker doors, grabrails, pillars, fiddles and so on.

Standard sizes of berths, seats, galleys and a host of other components are in the *Boat Data Book* (see the Bibliography).

12

THE DECK PLAN

HOW TO SET OUT THE DRAWING

For this or any other comparable job the plan view of the whole of the deck, port and starboard sides, is needed. A quick technique is to trace the centreline and deck line – that is the sheer line in plan view – off the lines plan on a piece of tracing paper. Turn the tracing paper over, align its centreline with the one on the lines plan, and again trace the outline of the deck. This gives the port and starboard sides.

Some people work using one side only, usually the port side, but for beginners there are advantages in having both sides because the details are not always identical. For instance there may be an anchor on one side of the foredeck only. Incidentally it will need chocks and eye-plates (or something similar) for its lashings.

All known relevant information is drawn on the deck plan, which usually includes the size and location of the rudder head, mast and chainplates, as well as the cabin top and cockpit. All the deck fittings are drawn in to scale. It helps to have a handful of photos taken of similar craft, showing their deck fittings.

SOME PRACTICAL DIFFERENCES BETWEEN CRUISERS & RACERS

On a racing boat the deck fittings are made as light and small as feasible and, where possible, things are not fitted if they can be left off, even at some inconvenience. This is to save weight, and perhaps to save money too. It also gives a clear uncluttered deck space which is so important for these craft.

An example of these extreme measures is that most racing yachts are either kept ashore or in a marina and so need no arrangements for anchoring, except perhaps to a small kedge in a calm. In such a situation the warp can be protected by a rag at the deck edge, and led to the single mooring cleat or bollard.

Cruising yachts, however, need to be able to moor up in rough harbours, and the more deep-sea orientated the yacht, the larger and tougher all the deck equipment must be. There will normally be pairs of fairleads and cleats at both bow and stern. Craft over 12 metres (40 feet) have extra cleats and fairleads amidships.

Fairleads have to be as close to the topsides as possible otherwise the deck gets chafed. Sometimes it is necessary to have half-round section lengths of metal strip to protect the deck. Cleats must not be so close to the fairleads that it is difficult to get the rope round nor so far that the crew have to move to secure the rope once it is through the fairlead.

FOREHATCHES NEED SOME THOUGHT

A forehatch is fitted on all but the smallest yachts and it is usually over the aft end of the fore cabin with the hinges across its forward edge. It should be located on the cabin top to keep it clear of any water sloshing on deck in moderate weather. If this cannot be arranged, then the hatch should be on a raised coaming, unless it is on a racing yacht where it is ideally flush with the deck. This allows the foredeck hand to tramp about on it without noticing it is there.

It will need non-slip strips glued on top, as will other deck areas where the crew work. These are drawn with cross-hatching or some similar technique. The dimensions of the hatch are taken from a catalogue / website. There is the size of the hole through the deck, which should not be much less than 600 x 600mm (2 x 2ft) as this allows a person in oilskins to scramble through. The dimensions to the outside of the frame are shown on the plan, as well as the location of the hinges.

ENSURING ADEQUATE VENTILATION

Many yachts have only one or two small vents, often mushroom or 'inverted plate' vents. But if voyaging to hot latitudes is planned there should be 450mm (18 inches) high cowls with water boxes. High vents need guard-bars over their tops to prevent sheets whipping them overboard, so it makes sense to group cowls together, and locate them above a bulkhead so they feed air to both cabins just beneath.

Vents are needed over the galley stove and in the toilet compartment. To help deal with this sort of problem some designers dot in the bulkheads on the deck plan. Others just lay the deck plan drawn on tracing paper over the cabin plan to get the precise location of the vents, tank filler pipes, the chain pipe and so on.

Any vent less than 100mm (4 inches) in diameter is ineffective and vents without water boxes are no good even for inshore cruising. Plastic mushroom vents are flimsy and best avoided.

If the yacht is going to be short of daylight inside, vents which double as deadlights are used.

COACHROOF DETAILS

The sliding hatch into the cabin is drawn in, with the hole size and the maximum dimensions noted. Forward of this hatch there is a 'garage' which is a cover like an inverted tray with the aft lip cut away. When the hatch is slid forward it disappears into the 'garage', which is designed to prevent water getting in at the front and sides.

A SAFE WORKING DECK

Handrails are shown on the plan with a note of their material and fastenings. On a cruiser they must be close enough so that the crew can go all over the deck and never be out of arms' reach of one. The stanchions are drawn in at 1.8m (6 foot) centres or less.

At the bow and stern there will be pulpits and these often need a special plan, or they can be drawn in elevation in a corner of the deck plan. They are made from 25mm (1 inch) diameter stainless tubing and need triangulating, especially on deep-sea cruisers.

The type and size of the winches is taken from a catalogue. One wants to stick to one make, and where possible a limited number of sizes of winch to reduce the number of spares needed.

Generally cheap mass-produced yachts have winches one or two sizes too small, so this is where it is often best not to copy other craft. In this connection an experienced yachtsman (who may know little about design) can be a great help when drawing the deck plan, or the cabin plan for that matter, as time at sea is a good teacher.

The construction plan should show all the areas of deck which are to be reinforced for items like the winches, cleats and anything else which will be stressed.

When locating deck fittings the aim should be to make walking about and working easy. This means keeping the side decks as clear as possible, and grouping foredeck fittings so that there is plenty of clear space. Envisage the boat pounding to windward well heeled; this means the foredeck crew need something to lodge their feet against such as a toerail or handrail running fore and aft down the centreline of the foredeck. On a large yacht two or even three such rails will be needed.

All round the outside of the deck there is normally a toerail which is often an aluminium extrusion. It may be an angle-bar with slots to take the shackles which hold spinnaker sheet blocks and such like, or it may be a special alloy section which extends over the edge of the deck to protect it.

Water and fuel filler caps are regularly located on the sidedeck where they are vulnerable. Seawater constantly running along the decks may seep into the tanks, so it is best to have these filler caps on top of a cockpit coaming.

MAKING THE JOB A LITTLE EASIER

Experienced designers save time by making up stencils for fittings which have to be drawn often. They take a piece of thin Perspex and draw the fitting in question on it to the required scale. The shape is cut out and the scale marked beside the fitting 'template'. This has the added advantage that the fitting can be tried out at different sites till the best one is found.

A stencil bought in an art shop or from a draughtsman's suppliers which has circles of different sizes is handy for drawing filler caps and bollards, and can be used for stanchion base flanges even though the shape may not be quite right.

Sometimes an item cannot be found in a catalogue, so its size is unknown, yet it still has to be put on the deck plan. Some designers just mark a cross for the centre of the fitting with an explanatory note beside it. A better technique is to draw the fitting substantially over-large and make a note that it is NTS (Not To Scale). Then there will be no chance that another deck fitting will be drawn too close.

COMPASSES & COCKPITS

Siting the compass is often difficult and it is common to have one each side. Even then the precise position may be hard to fix because the helmsman has to be able to see the course when the yacht is heeled or rolling, in heavy rain or when the cockpit is crowded. The same applies to the instruments.

A popular way of getting around this is to make a mock-up of the cockpit or even of the whole deck plan. The designer can often learn a vast amount by using boxes or benches for cockpit seats, trays for hatches, chalked lines on the floor, sticks held by duct tape for the tiller and also for the stanchions and boom gallows. If building a normal fibreglass yacht, the location of the fittings can be decided on the deck / cabin-top plug.

Alternatively the deck plan can include section drawings, especially in way of the cockpit, looking forward and aft. Some designers even do the deck plan in elevation, especially if they are worried about the view forward from the cockpit. If all else fails, arrange for the boatbuilder to supply buoyant cushions to raise the helmsman and improve his view forward. He will also need foot rests, which should be adjustable for different sized people.

CHECKING THE DESIGN

Once the deck plan, or any other plan for that matter, has been completed, it is a good idea to take it on board a comparable craft to see if anything has been forgotten. This ploy occasionally shows better locations for some items. It may remind the designer of the need for wedges or plinths under the base flanges for stanchions to make them upright even though the deck is cambered. Chocks, often of hardwood, are also needed under permanent boom gallows and deck sheaves and perhaps for the compass, and so on, to ensure they are correctly aligned.

13

COSTING A DESIGN

UNDERLYING PRINCIPLES

In this chapter the word 'unit' is used instead of pounds, dollars, francs or any other currency. In this way anyone in any country, or costing in any currency, can use the methods described here without difficulty and without converting from one currency to another.

There are several ways of predicting the cost of a boat and, as most involve a mixture of science and art, it is sensible to use more than one technique to cross-check the final results. Even so, the answer must always be treated with caution.

Wise estimators allow a 'fudge factor' or add on a percentage to deal with unforeseen occurrences such as delays in the arrival of important parts, or breakages during construction.

If in doubt certainly add at least 3%; 6% is not excessive.

BY DIRECT COMPARISON

The first method is to compare the proposed craft with one recently built or on offer from a firm which sells new yachts, making sure that the two craft are closely similar. In practice it may be hard to find a vessel nearly identical, so allowances have to be made for different sizes of engine, a different number of sails, different instruments and so on.

It is easy to discover the cost of the engine (or sails or instruments) fitted to the yacht which is being compared and the

engine proposed for the boat being costed by contacting the engine makers. If the engine in the yacht being discussed costs a thousand units more than the engine in the rival craft already on the market, then the new boat will cost that much extra.

What upsets this simple comparison is differences in production runs. An amateur designer will probably be trying to price a single yacht, and the only ones available for comparison will probably be series-buillt boats which have come from a factory turning out lots of identical vessels. As a rough rule a 'one-off special' will probably cost between one-and-a-quarter and one-and-a-half times as much as a 'mass-produced' yacht.

RULE OF THUMB

Costing One-Off Specials:

As a rough rule a one-off special will cost between 125% and 150% as much as a comparable mass-produced yacht.

THE FIVE FOOT ROUGH RULE

A useful guide when estimating the cost is to use the 'Five Foot Rough Rule' which I invented some years ago, and which may need modifying from time to time. This 'Rule' states that the price of a yacht more or less doubles with every increase of roughly 1.5m (5 feet) in overall length.

There are snags here: for instance it is essential to compare like with like. It is no good saying that because a lightly built dayboat with one suit of sails and a single pair of winches, plus a small outboard engine costs 20,000 units, then a cruiser 1.5m (5 feet) longer, with an inboard, three times as many sails, elaborate deck and cabin fittings will cost 40,000 units.

It is just as important to compare building quality, complexity, final use and standards of finish, like for like. But the rule can be wonderfully handy, especially as it can be used upwards or downwards, adding or subtracting 1.5 or 3m (5 or 10 feet).

The rule should not be used for craft over about 14m (say 45 feet) or under 7.5m (about 25 feet).

RULE OF THUMB

Five Foot Rough Rule:

The price of a yacht between 7.5m (25 feet) & 14m (45 feet) LOA roughly doubles for every increase of 1.5m (5 feet) in overall length.

| | **Yacht 1** | **Yacht 2** |

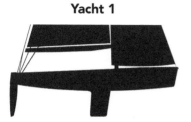

	Yacht 1	Yacht 2
Length	10m	11.5m
Cost	£Z	2 x £Z

ADDING UP ALL THE PRICES

The way many professionals cost proposed vessels of any sort involves a lengthy and detailed investigation. Each part (all the materials, paint, fastenings, the engine, spars, the cooker, everything) will be looked up in catalogues / websites and the prices put down on reams of spreadsheets. All these prices are then added up.

Next, the number of man-hours for each part of the building process will be put down, stage by stage. This list will commence probably with laying down the building base, setting up the mould(s)... and so on to the launching and trials, followed by rectifying the faults found during tests at sea.

COSTS & OVERHEADS – THE BOATBUILDER'S PERSPECTIVE

The builder will know the cost per man-hour he is using currently, so he simply adds up all the hours and multiplies the total by the hourly rate. This gives the cost of the labour which is then added to the cost of materials and equipment – and that is the basic cost of the yacht.

A professional builder will also add on a figure for profit, and probably a contingency figure to deal with unexpected crises. He will also add appropriate taxes.

For an amateur-built boat there is no labour figure, in theory. In practice many amateur builders call in one or two specialists, perhaps to install the engine, or to fit the electronics. Firms doing this sort of work will quote for the job without charging for their quotation, and their work is added to the cost of the materials.

There is also the cost of electricity, rent of the building shed, water and so on. These have to be added to the cost of an amateur-built boat on some occasions. So far as the professional builder is concerned, these expenses are in the 'labour' figure, and are called 'on-costs' or 'overheads'.

When a professional firm is doing its estimating, it uses a 'labour' figure which is in two parts. The first component is what is actually paid to the man doing the physical work. Let's say he gets paid 6 units per hour. The firm's on-costs may be 150% – to cover the office expenses, the manager's and foremen's salaries, the cost of heating and lighting and so on. So the firm charges its labour out at 6 units to cover the wages plus 1.5 x 6 units to pay for the on-costs, Making the total labour figure 2.5 x 6 units, namely 15 units.

There are often several 'wage rates' or 'labour rates' in a boatyard. Skilled craftsmen are generally charged out at a higher rate than semi-skilled or unskilled boatbuilders because they are paid more and make more demands on the back-up services the yard provides. This variation can affect the cost of the craft a lot, especially if she is built by a simple technique which does not require a high input of expert labour. Fibreglass boats tend to be relatively cheap as a result of this factor.

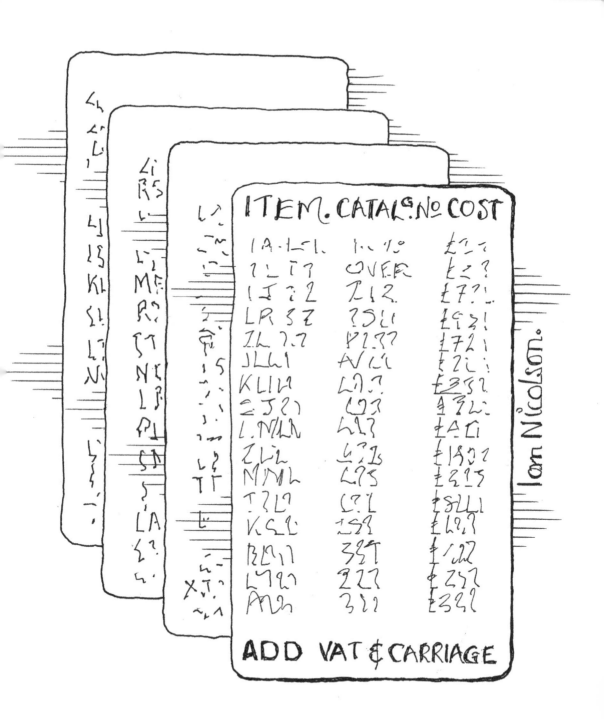

Costing a boat involves listing all the gear. To do this, work through chandlers' websites / catalogues and, in this way, every needed part will be included. It's worth remembering about the famous naval architect who once designed a cruiser and omitted to include a loo in the plans and specification!

THE THREE TIMES MATERIALS RULE

Anyone who wants a rough idea of how much a boat will cost can use the traditional 'three times materials' rule of thumb. This says that the final cost of a one-off yacht is very roughly three times the cost of the materials, parts and fittings.

This is handy for amateurs who cannot go to a boatyard for an estimate because the design is far from complete. It is a lengthy but simple job collecting all the costs of all the items and adding them together.

First write to major chandlers for their catalogues and price lists, then write to or e-mail engine makers, spar makers, sailmakers and so on.

A good approach is to start at the front of the catalogue and work right through

> *RULE OF THUMB*
>
> **Three Times Materials Rule:**
> The final cost of a one-off yacht will be roughly three times the cost of the materials, parts and fittings.

it. In this way the risk of missing anything out is small. As chandlers don't sell sails, timber or specially-made fittings, the designer has to be alert and check his list of costed materials and components with the preliminary design and specification.

Once the list has been completed a contingency reserve figure should be added to allow for items which may have been missed, for inflation and so on. Delivery costs also have to be added.

SOME OF THE SPECIAL FACTORS TO AFFECT A COSTING

There are a dozen special factors which affect estimating. A boatyard may be short of work – if so it may reduce its labour nominal rate, or work for a tiny profit, just to keep their personnel employed. On the opposite tack, if the method of construction required by the designer is novel or difficult, the estimator will be worried and probably add a great many units to deal with time spent learning the new techniques, materials wasted due to mistakes with new processes and so on.

Designers keep costs down by sticking to well tried methods and minimising the number, size and complexity of fittings. This is also achieved by specifying standard mass-produced items rather than parts made specially. For example, a cheap cabin top is one which is short, low and narrow. It also has the minimum number and size of windows, of a standard mass-produced type. Equally it may be possible to use U-bolts instead of specially fabricated chainplates.

PART 3

MAKING ALTERATIONS TO A DESIGN

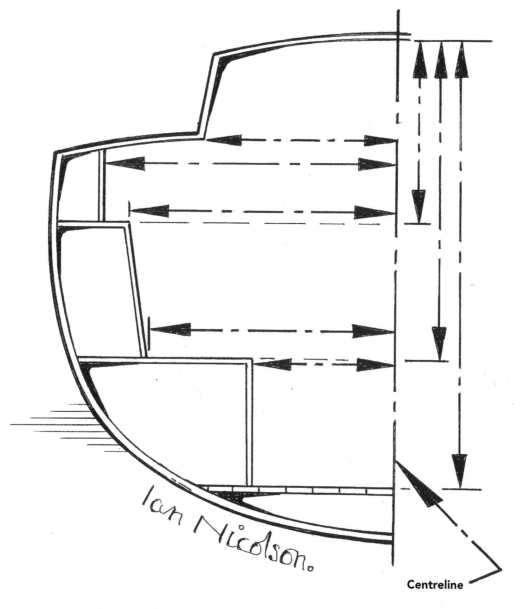

Centreline

*When changing the cabin plan a great many dimensions have to be taken of the available space.
The centreline of the yacht has first to be drawn along the cabin sole and at the deckhead.*

14

CHANGING THE SAIL PLAN

If a drawing of the sail plan already exists, then changing it is mostly a matter of altering the shape and size of the spars, sails and rigging, while making sure that the centre of effort remains in the correct place.

When there is no sail plan, the first job is to take a lot of measurements off the yacht and make a scaled drawing. To do this a thin wire or line is stretched very tightly from bow to stern. Equal intervals, which are typically 1.2 metres or 4 feet, are marked with chalk on the deck below the line, or with coloured ink on the line.

Exactly opposite these marks, the distance from the deck edge down to the Load Waterline is measured. This gives a series of points above a straight horizontal line which is drawn on the new plan to represent the Load Waterline. Joining these points with a smooth curve results in the sheer in elevation. The profile of the bow and stern is established by taking more measurements off the yacht or off a photo taken from dead abeam.

The length, height and location of the cabin top(s) are taken and marked on the drawing. The diameters and locations of the spars and chainplates are noted, as well as the points where the forestay, backstay and other rigging meets the deck. A careful draughtsman measures from both the bow and the stern to double check all dimensions.

To get the rake of the mast, a thin line is hoisted on the main halyard on a windless day to act as a plumb line. A weight is tied to the line just above the deck. The distance is then measured from the middle of the weight to the back of the mast, from which the mast bend has been removed. The distance from the main halyard sheave to the deck is also measured, and this gives the mast height to the middle of the sheave, and hence the mast can be drawn in correctly raked.

The size of the fore triangle is measured using the genoa halyard to pull the tape up the mast as high as it will go. Ideally someone will go up the mast to measure from the halyard shackle to the front of the mast, following the line of the forestay. The process is repeated aft of the mast using the main halyard.

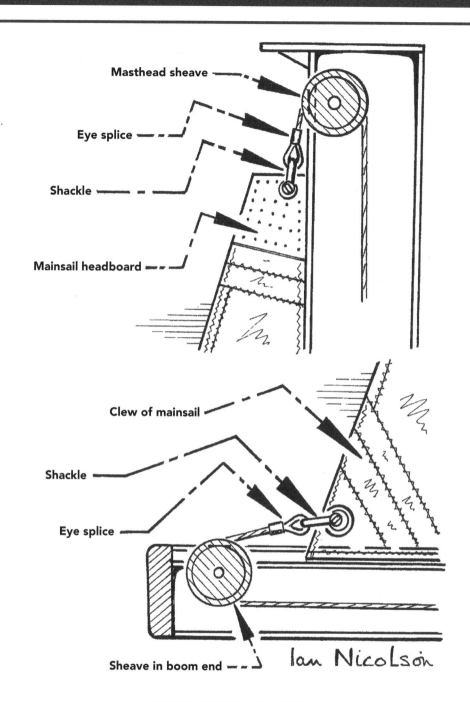

Masthead sheave

Eye splice

Shackle

Mainsail headboard

Clew of mainsail

Shackle

Eye splice

Sheave in boom end

Ian Nicolson

VIEWS LOOKING TO PORT

SAIL AND SPAR DESIGN

The mainsail cannot extend right to the top of the mast, or to the end of the boom. The reasons for this are shown on this drawing, which shows that there must be length for the rigging and the sheaves.

Measurements are taken along the top of the boom, the boom diameter (or diameters if this varies) and the height of the top of the boom off the deck. Other dimensions needed are: outer end of the boom to the outhaul sheave axle, sheave diameter, and outhaul length in its normal position.

One also needs to know the distance from the mainsail tack pin to the back of the mast, and from the top of the boom to the gate on the luff track. The location and length of the sheet lead tracks is also marked on the plan, with the height and location of the winches, sheet lead blocks and so on.

With all these dimensions a sail plan can be drawn. To be safe a lot of dimensions are taken to give a cross-check. For instance the distance from the main halyard sheave to the end of the boom is noted, as well as the distance down the forestay from the headsail halyard sheave(s) to the top of the deck where the forestay terminates. Other check dimensions include the distance in line with the top of the boom, from the end of the boom to the backstay.

Ideally all the sails are measured by a sailmaker on a sail loft floor. He will know how to pull the sails just tight enough to get the correct dimensions.

All the sails are drawn in on the plan, which incidentally works as a further check for accuracy. The sail plan will still be less than 100% precise, but it should be adequate for most purposes.

Changes to a sail plan are often made in consultation with a specialist mast maker and a sailmaker. These people handle lots of their products annually whereas the busiest designer does not work on many different craft in a year.

REDUCING THE SAIL AREA

If a boat is over-canvassed, or heels too much, or is to be cut down from a racer to a cruiser, then the sail area is reduced forward and aft in such a way that the final centre of effort isn't moved forwards or aft.

The easy approach is to cut off the leech of the mainsail and use the smaller headsails in the yacht's wardrobe. A new mainsail will be needed for racing but for cruising the cut-down sails are often satisfactory provided the job is done expertly.

If 15% or less of the mainsail foot is cut off, the boom seldom needs altering. In any case cropping a piece off one or other of the boom ends is seldom difficult.

However the mainsheet attachment point may need shifting. This sheet should be no more than 25 degrees off the vertical when the boom is amidships.

It may be necessary to move the bottom of the forestay aft. To do this a new deck fitting will have to be designed and made, or the existing one altered to suit its new location. It will also need an under-deck structure to take the upward pull of the forestay down to the stem. Alternatively a strong beam may be fitted on a small boat to carry this forestay tension out to the topsides. For craft under 9m (30 feet) a U-bolt above deck, instead of a special

---

RULE OF THUMB

Mainsheet:

The mainsheet should be no more than 25° off the vertical when the boom is amidships.

---

deck fitting, can often be used to take the bottom of the forestay.

Where a more elaborate reduction in the sail plan is needed, the mast(s) as well as the boom(s) will have to be shortened. This is not a dramatic job for an experienced shipwright regardless of whether the spars are made of wood or aluminium alloy. The crosstrees are almost certainly going to have to be moved, as they should roughly divide up the mast into even sections.

The angle between the shrouds and the sides of the mast should be 15 degrees or more for an ocean cruiser. On a coastal cruiser an angle of 13 degrees is widely held to be a safe minimum, but on a racing yacht the angle is often less. How much less depends on such factors as the mast diameters and its wall thickness, so it is best to consult a specialist mast maker.

RULE OF THUMB

Angle of Shrouds to Mast:

Angle between shrouds and the sides of the mast

Safe Minima:

Ocean Cruiser: 15°

Coastal Cruiser: 13°

Moving crosstrees on an aluminium or wood mast is normally straightforward but if a number of other changes are needed and the mast has been damaged, or is something like 25 years old, it may be sensible to go for a new one. Often the designer simply draws out what is required and leaves the mast maker to organise the alterations or replacement in detail.

Where the sails are being cut down to make a cruiser into a motor-sailer (to suit an ageing owner), or for deep-sea sailing with a limited crew, extra and deeper reefs may need adding to the drawing. Likewise single-line reefing, lazy-jacks, full or three-quarter battens may also be added for safer sailing. Occasionally the line of reefs is angled up about 10 degrees towards the aft end to keep the boom end out of the water when the boat is heavily heeled.

Cutting off the leech of a mainsail tends to have little influence on the balance of a yacht, or on the amount the tiller tugs at the helmsman's hand. So if the helm feels fine, but the total sail area is too high, it makes sense to reduce the mainsail foot length first without changing the headsails or the location of the forestay. During sea trials it may be discovered that the helm balance is still good so there is no need to shift the forestay.

INCREASING THE SAIL AREA

If a yacht is slow, or a mistake has been made in her original sail plan, or if she's going to race, it may be necessary to work in extra sail area.

The centre of effort must not be moved forward or aft in theory, but in practice it is often possible to increase the area forward by adding a moderate bowsprit without upsetting the balance.

It is seldom easy to improve the sail plan without going for a higher mast. However,

plenty of masts can be lengthened more cheaply than the cost of an entirely new spar. So the designer just draws out his requirements, and discusses with the spar maker whether the existing tube is to be cut and lengthened with an inserted piece (assuming it is of light alloy) or an entirely new mast is to be made. Wood masts are often easy to lengthen by gluing on a new piece with a long, carefully sloped scarf.

Here, as always, much depends on

the level of determination and finance available. If an important series of races is to be won and cash is plentiful then a new mast will be ordered automatically. At club cruiser-racer level the existing mast will be altered and much money saved, especially if a lot of the work is to be done by the owner and crew.

Extra area can be added to the leech of the mainsail by having longer or full length battens, but this may only work if the mast is moved forward to preserve the helm balance. Sometimes when there is a problem getting added area, and a bowsprit is not favoured, or only a short bowsprit can be fitted, the answer may be to move the mainmast forward and fit a mizzen.

Designing a ketch or yawl tends to be tricky because the sum of the sail areas is so often too far aft. Moving the mainmast forward is seldom an easy job and care is needed otherwise its weight, located towards the bow, tends to make the yacht wetter and the bow slower to rise to on-coming seas. Clever designers compensate by using specialist light standing and running rigging, as well as the lightest spars commensurate with reliability.

Locating a mizzen is not easy as it must not spoil the helmsman's view, or be too obtrusive so far as the crew are concerned. It has to stand on a strong structure so extra scantlings have to be worked in to support the mast and chainplates. The supporting components should be designed to take three times the maximum expected mast compression.

ALTERING WEATHER HELM

A yacht should have a small amount of weather helm when going to windward. Between 2 and 5 degrees is sometimes quoted. A lot of weather helm makes for tiring steering and may damage the steering components.

It is possible to reduce weather helm by various techniques, but adjusting the sail plan is often the most effective. However, even major changes will not work if the yacht's hull shape is too unbalanced.

Sea trials should be carried out to discover what effect any changes that have been made have on weather helm. A temporary bowsprit may be rigged in moderate weather with a headsail set from the fore end, before a permanent bowsprit is made.

The yacht should also be sailed with a reefed mainsail and full genoa, then with no mainsail, to see how the helm feels under headsail alone.

Lee helm, which requires the tiller to be held to leeward to keep a straight course, is rare but dangerous. It is cured by moving the sail plan further aft.

SOME SERIOUS CONSEQUENCES OF MOVING A MAST

Moving the mast forward is expensive but is sometimes the only way to reduce a dreadful and vicious tendency to luff hard all the time. A new structure to support the mast and chainplates will have to be fitted to suit the new mast position.

A deck-stepped mast will have a pillar or bulkhead or both under the deck. Moving a pillar is sometimes easy provided it is not secured to a bulkhead. However the new area where the bottom of the pillar rests may need beefing up to take the down-thrust of the mast. As a rough guide, this should be assumed to be about the same as the weight of the yacht, and a factor of safety of at least three is needed. To allow for unknown weaknesses and unexpected problems, it is good practice to specify new scantlings which are longer, wider and thicker than the original set.

Where a mast is being moved a short distance, a deep strong well-secured bracket may be fitted on an existing bulkhead. As the stress generated by a sudden load is twice or more than that applied gradually, and as the stress on a mast is often a jolt, factors of safety around six or more make sense, especially for craft used far offshore.

A vertical support under the middle of the mast is almost always the ideal approach. Fitting a bracket on the face of a bulkhead or pillar has to be considered a compromise, though it may be necessary to avoid cluttering the accommodation too much.

A fibreglass deck under a deck-stepped mast must be solid. If there is a soft core in a sandwich deck, this must be routed out and hardwood or marine ply or even metal must be substituted and glassed over. The designer needs the original construction plans to see exactly how the deck is built, but even then has to assume that the plans may not have been followed exactly. The deck may first be drilled to see if there is something solid under the top layer of fibreglass.

To be safe it is best to take off the top layer of fibreglass and check beneath. The new mast step fitting will cover the signs of excavation and subsequent repair. Likewise it is essential to check the status of the structure at the new location of each chainplate. Other things being equal, the chainplates will need to be moved the same distance as the mast. However, if the distance the boom swings forward isn't critical (and only on a racing boat does it really matter), then the aft set

of chainplates may not need moving if the mast is only shifted a short distance. Sometimes the original middle chainplates can be used as the aft ones once the mast has been moved forward.

Chainplates have an upward pull similar to the compression load on the mast, so on each side of the yacht all the components which hold the lower ends of the shrouds should be strong enough together to support the weight of the yacht. There must be a factor of safety of at least two, even for an inshore yacht, and four for a long range cruiser.

However the thickness of the chainplates and the size of the holes for the toggles on the rigging screws, as well as the distance down from the top of each chainplate to its hole, must all fit the chainplate toggles which in turn depend on the diameter of the shrouds.

Here, as so often, the best way forward is to buy the dominant component – the rigging terminal – and make everything to which it links fit exactly. Because many mass-produced craft are lightly built to save money, replacements should be stronger. It is no bad thing to aim for an increase in thickness and width of 33% everywhere.

If, as is common, the chainplates are bolted to ply knees or bulkheads, these should certainly be made stronger than those previously used, especially if the yacht is one of a series-build low-price type. Extra under deck stiffening in way of the chainplates also makes sense.

Even if there were no deck plates in way of the chainplates they should be included in the new design. Normally they are made to fit smoothly round the chainplates, and are pulled down onto the deck over copious bedding with four screws or bolts. They prevent water seeping through the deck beside the chainplates and also resist lateral forces on the chainplates.

15

DESIGNING AN ALTERATION

PROPER DRAWINGS ARE IMPORTANT

When an existing boat has to be altered, the work should never be done without proper drawings. If there is no detailed plan, unexpected crises will occur. Predicted finishing times and costs will overrun, ordering material will be haphazard and something vital such as special bolts will be forgotten until urgently needed.

The change may involve fitting a new engine or rig, or a new layout for the deck or the saloon. The first move is always the same, namely to get hold of a set of prints from the original plans.

When these are unobtainable it is sometimes possible to get hold of a sales leaflet with a simple sail plan, cabin layout and elevation plan. Though these pictures will be to a small scale and often short of detail, they can be 'blown up' by a drawing office supplier using a sophisticated plan printer or copying machine.

This enlargement has to be done carefully so that the prints obtained are accurately to scale. In practice this sometimes means doing three or even five prints till the overall length of the vessel on the plan is exactly right to the selected scale.

Where no drawings are available a lot of dimensions are taken off the boat and put down on paper or into a computer. These sizes are in the form of a grid with lines running fore and aft, also athwartships. The job is easiest when the deck layout is being altered because the sheer can normally be ignored, at least initially, and the work carried out on the basis that the deck is flat and horizontal.

A NEW DECK LAYOUT

The first job for a new deck arrangement is to draw the centreline on the boat, using chalk. Some people prefer to use masking tape in short lengths with one edge carefully stuck down along the centreline. It is important to mark which edge is being used to avoid future confusion. Here, as elsewhere in this whole operation, different coloured marker pens which are erasable are a considerable help.

Next the centreline is marked off at suitable intervals – typically one metre or three feet. Each mark is numbered from the bow to establish the overall length. At each mark the exact width of the boat is measured, taking care to align the tape precisely athwartships. To be certain that the width is measured exactly at right angles to the centreline a 1.5m (5 foot) T-square or set square is used. These dimensions are put onto a rough plan for transferring to a neat properly-scaled plan later.

Next all the features on the deck are marked on the rough plan, including the location and size of the cockpit and mast base, the stanchion bases, sheet winches' base dimensions and location and so on.

The cabin and cockpit coaming outlines are obtained by measuring from the centreline.

The designer draws an accurate plan using all the collected dimensions. A piece of tracing paper is laid over the plan of the existing layout. The outline of the hull is traced, together with the cabin coamings, the cockpit size and location and any other feature which is not being changed. In passing, it may be that the cabin top and cockpit are to be altered, in which case they need to be drawn in accurately at their new locations.

The new position of each item of gear is drawn in using suitable templates for items like winches and hatches. These templates are supplied by the manufacturers and are often available to different scales, so before beginning the job it makes sense to get the templates and see what scales are available, or alternatively make your own.

A NEW CABIN PLAN

When altering a cabin it is essential to start off with accurate drawings of the yacht in plan, elevation and section view. As a minimum there must be a section at each end of the cabin, and one in the middle is often an asset. In theory altering a deck plan does not require elevation and section views unless the changes involve a new cockpit or cabin top.

In practice cynicism is the best basis for good draughtsmanship and all three views should be drawn out before and after the changes. Paper and drawing time is almost always cheaper than small mistakes – and far less costly than major errors when tools are being used on the boat. When making a plan of an existing cabin with a view to making alterations it is normal to make the bulkheads at each end of the cabin the basic grid lines in plan view. In the same way, it is a good idea to use existing furniture sides, ends and tops as the basis for the grid.

For instance, after the centreline has been marked on the sole and maybe along the deckhead as well, the distances off the centre of each berth front, berth back, galley front and back and so on are measured and noted on the rough sketch. However, it should not be assumed that any of the furniture fronts run parallel with the centreline, so the distance off the centreline at each end of each component must be measured.

It is a big help to have someone to hold one end of the long tape measure. Failing this, some designers mark long light stiff wood battens in centimetres or inches and use them for measuring as they can be used one-handed. Duct tape is also handy for holding the end of a long tape. Various measuring devices are available such as folding zig-zag wood rules which save making up special battens.

When making the drawings based on all the dimensions measured off the boat, it will be found that some points on a curve do not 'fair up'. For instance, when drawing the line of the inside of the hull at cabin sole level, there will typically be points every 600mm (2 feet) along the inside of the hull. On the neat plan an attempt is made to draw a fair curved line joining up these points. In practice not all the points will lie on the final line. It is, instead, what might be called an 'average' of the measurements, and it is almost always accurate enough for the job in hand.

What look like errors in measuring are common; they should be quite small, and typically will be of the order of a centimetre or two – around 1/2 inch. Because precise measurement is often nearly impossible, the aim is to go for an accuracy which is about the same order. This means that when planning a new layout, this level of inaccuracy should be assumed to exist. So nothing is designed precisely to the nearest centimetre – or half inch. Berths should be made a couple of centimetres (say 3/4 inch), wider than standard and an extra 25mm (or an inch) is left each side of a standard cooker, to be sure it fits when it is installed in the altered cabin.

SOME USEFUL POINTERS

Before designing a new plan it is sensible to study the work of the best designers. Taking photos from different angles of top quality boats helps. Before putting in new furniture or any fittings, sizes are needed for such things as berths and bookcases. These and a host of other essential information are listed in *Boat Data Book* (see the Bibliography).

The thickness of materials has to be included, and should be noted in the specification. This is a detailed list of all the work to be done and the things used for this work. One aim should be to use the same sizes and thicknesses throughout, to save money and time. For instance, if hinges are bought wholesale in boxes of ten, they will all need the same size of screw. So there is an obvious need to standardise.

THE BASIC RULES FOR DESIGNING ALTERATIONS

These basic rules apply to all alterations:

1. What works in harbour may be dreadful at sea, especially when the vessel is heeled or rolling. The clever designer thinks about where the crews' feet will end up on the edges of the cabin sole. Everyone needs grabrails everywhere at sea.

2. Arrangements, materials, fittings and bright ideas which suit houses and caravans are often a disaster offshore because the sea is not stationary or flat or benign.

3. It often makes sense to copy another successful small craft. However, many current products are built down to a price and items on board are used because they are cheap, not necessary because they are satisfactory. For instance, door and locker hinges are seldom stout enough to stand up to deep-sea conditions.

PART 4:
TO USE COMPUTER-AIDED DESIGN OR NOT?

Ian Nicolson

*As a yacht design can take many hundreds of hours to complete,
a comfortable chair is needed. Also a table at the correct height.*

16
COMPUTER-AIDED DESIGN

ARGUMENTS FOR & AGAINST CAD

The advantages of using a computer for design are numerous. Calculations that are pretty tedious with a calculator can be completed so much more quickly. Computer-Aided Design (CAD) can eliminate much of the tedium of traditional design work and alterations which used to take hours can be done in minutes.

A beginner should learn and practise the old-fashioned way to understand what the computer is doing, and so be able to spot errors. Also a lot of the design stages have to be done by a human brain making compromises and decisions. A brief sketch followed by a thoughtful drawing gives the mind time and stimulus to sort out the best alternative.

For instance when the first calculation of the displacement is made it is frequently found that the design needs fining down or filling out. If the basic hull shape is just right it goes against the grain to change it. The clever thing may be to change the keel or skeg or both. The change may be made by lengthening, deepening or thickening, or changing all three dimensions, increasing some and decreasing others. The outline shape of these appendages may be changed at the same time.

This sort of alteration can be done on the computer, but pencil sketching is often the best way to start the alterations, as the nuances of each change can often be better appreciated.

THE KIT

A complete computer set-up is needed, including a printer which ideally should be able to handle A3-size paper to give a useful size printout. Alternatively, for anyone with the cash, a plotter capable of printing rolls of drawing many metres in length is the ideal.

The software required is supplied by specialist firms who advertise in a magazine called the *Naval Architect*, available from the Royal Institute of Naval Architects: www.rina.org.uk.

PRINTING A DIGITAL FILE

Access to a big printer is needed as soon as the beginner in designing becomes a serious practitioner. These printers are expensive but some draughtsman's supply shops will run prints from a digital file. Many architectural or engineering design offices are prepared to help for a fee, provided their facilities match the designer's.

THE SOFTWARE NEEDED

It is essential to get small craft software which suits the designer's own computer. The package will include what might be called 'Basic Designs'. These are hull forms which are typical shapes, from which the designer develops his own choice of size, proportions, keel shapes, rudder size and shape and so on.

The designer starts with the preliminary shape on his screen and uses the mouse to nudge the lines around to suit his new design. Some programs start with a rectangle or cylinder but the beginner should avoid these.

The software package should include comprehensive instructions plus a helpline to deal with difficulties. Knowledge, skill, plus ample study is still needed because few of the packages available will explain things like the percentage of the total vessel's weight which is in the keel. For a racing yacht it will be around 60% and for a heavily-engined cruiser the figure will be about 35%.

Using a computer, changes in hull shape can be made quickly. If the Load Water Line is filled out, the buttocks and sections move out simultaneously and in precise accord. This is in strong contrast to the traditional pencil-pushing situation where lines have to be rubbed out and new ones plotted then drawn in. This precise work takes time and needs care because alterations to one line often result in ugly, knobbly, unacceptable curves in a different plane. When this occurs the amount the first line was moved may have to be increased or decreased, till a sweet curvaceous form is achieved.

Computer-Aided Design enables the hull shape to be viewed from different angles in perspective. The traditionalist will say that a half- or full-model gives more information, and it is hard to deny this. But it may take weeks of work to make a good accurate model.

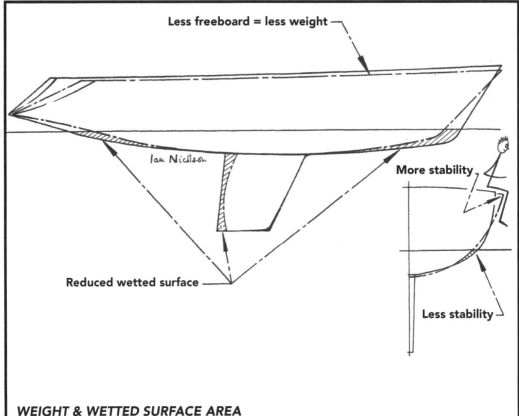

Less freeboard = less weight

More stability

Reduced wetted surface

Less stability

Ian Nicolson

WEIGHT & WETTED SURFACE AREA

Boats which are fast are light and also have a small wetted surface. The elevation shows techniques for reducing weight and wetted surface. The steeper slope of the transom reduces the area of the topsides port and starboard, so it also reduces wind resistance. For speed to windward, racing yachts have the crew sitting on the windward deck edge. The further outboard they are, the more effective their weight. By reducing the waterline beam the hull resistance is reduced.

COSTS ARE COMING DOWN

The cost of computers and their software, printers and associated components comes down steadily. Anyone who wants to explore this field should look at several software suppliers and get hold of their demonstration programs.

SOME REAL ADVANTAGES

Disks from component suppliers and manufacturers are also a help, as they allow boat equipment to be added to a plan accurately and swiftly. Of course there is still plenty of skill needed. The outline of the WC, for example, may seem to fit wonderfully well in the toilet compartment on the drawing in one view. But it needs locating and checking in all three views.

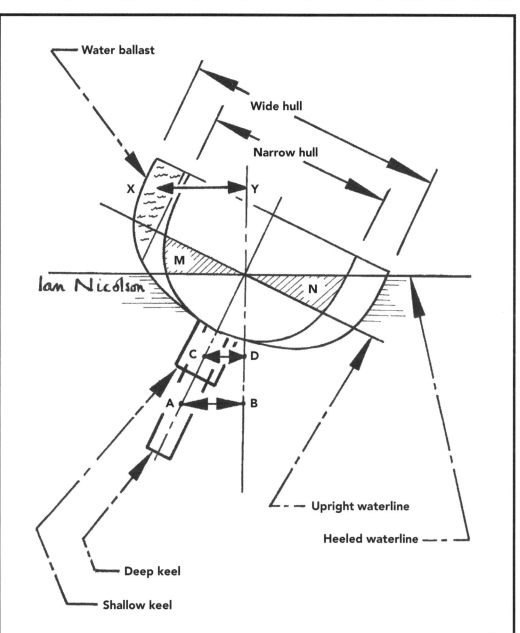

IMPROVING STABILITY

When a yacht heels a 'wedge' of the hull comes out of the water, and here it is the shaded area M. Another 'wedge', labelled N, gets immersed. As N is bigger than M there is a thrust of buoyancy upwards tending to push the boat upright and offset the pressure on the sails.

A wider boat will usually have bigger 'wedges' and hence more stability. Stability can also be improved by having water ballast tanks well outboard. The stability increase is:

The Weight of Water in the Tank x The Distance Outboard X-Y

Yet another way to increase the stability is to have a deeper keel. The leverage of the shallow keel is the distance C-D, whereas that of the deep keel is A-B. However A-B is a lot smaller than X-Y.

GLOSSARY

AFT Near, towards or at the stern of the vessel.

AMIDSHIPS Halfway, or roughly halfway between the very front and the very back of the boat.

ATHWARTSHIPS Across the vessel, that is at right angles to the central fore-and-aft line.

BACKSTAY A piece of rigging extending from the top of the mast to or near the stern. It prevents the mast falling forward and helps to keep the forestay tight.

BALLAST KEEL The heavy weight secured to the bottom of a sailing yacht to help her remain upright when pressed over by the wind. The keel also prevents the yacht from sliding sideways too much when going against the wind.

BASELINE A horizontal line established by the designer, either the designed waterline (DWL) or a horizontal line parallel to the DWL below the boat's profile as drawn on the plans.

BOW The front of a vessel.

BOWSPRIT A spar extending forward from the top of the stem to secure the bottom of the forestay further forward than the stem and hence increase the sail area forward of the mast.

BUMPKIN A spar comparable to a bowsprit but extending aft of the stern to take a backstay or the sheet of a mizzen sail, etc.

CABIN SOLE The 'floor' on which the crew walk. (See also 'Floor'.)

CABIN TOP An erection above the deck which typically extends from the cockpit to forward of the main mast. It was first brought in to give headroom in the cabin(s), but it also serves to get light down below because it has windows down the sides.

CLEAT A fitting which holds the end of a rope such as a halyard or sheet.

CLEW (OF A SAIL) The aft corner.

COAMING The side of a cabin top or cockpit.

COCKPIT The area where the helmsman steers the boat, and where the crew spend most of their time when on deck in cruising yachts. It is usually aft, and consists of a sunken well (box) with bench seats each side. Usually the design is such that water which gets into the cockpit drains back to the sea, hence the phrase 'A self-draining cockpit'.

COUNTER STERN A stern which extends well beyond a vessel's waterline.

DECKHOUSE An erection above the deck added to enhance the vessel's living space. It may form a cabin but is more often part of a cabin. It is usually just forward of the cockpit and has the cabin entrance at its aft end, with windows down the sides and across the ends. It is comparable to a short high cabin top.

DESIGNED WATERLINE (DWL) The straight horizontal line extending from where the stem is drawn down to the water level, aft to where the stern is drawn down to meet the water. The 'Designed Waterline' may be the same as the 'Load Waterline', but for various reasons it may be located slightly higher or lower.

DRAUGHT The immersed depth of a vessel.

ELEVATION The view from the side, usually the starboard side, because by tradition plans are drawn with the bow to the right.

FIDDLE A tiny wall round the edge of a cabin table or galley or chart table, to prevent objects from falling off when the yacht heels. Usually made of wood, a fiddle should be at least 50mm (2 inches) high but seldom is in production yachts.

FLOOR A structural component extending athwartships across the top of the inside of the keel, located below the cabin sole. It holds the sides and bottom of the vessel together and strengthens the bottoms of the frames. A floor sometimes takes the top end of one of the bolts which support the ballast keel.

FORESTAY A wire which extends from the stem or near the stem to the top or near the top of the mast, to hold the forward sails and also to prevent the mast falling backwards.

FORE HATCH An entrance to the cabin(s) located forward of the mast. It normally hinges up and has a transparent top to let light into the hull.

FORWARD Near, towards or at the front end of the vessel.

GENOA A sail set on the forestay, that is, forward of the mast. It is used for going to windward and reaching across the wind. It is made of Terylene or Dacron for cruising yachts and more exotic fibres for racing yachts. A yacht often carries more than one genoa, of different cloth weights to suit light and strong winds.

GUARDRAILS Horizontal wires which extend round the deck sides to prevent the crew falling overboard. (See also 'Pulpit'.)

HALYARD A rope for pulling a sail up a mast.

HATCH An entrance to the cabin. The main hatch is normally at the fore end of the cockpit and has a sliding top (lid).

HATCH GARAGE A cover like an inverted tray which prevents water from getting in at the fore end of the sliding hatch.

HEAD (OF A SAIL) The top corner.

INTERNAL BALLAST Weight inside the yacht stowed low down to take the place of or to supplement the ballast keel.

JIB A small genoa.

KEEL Structure extending below a ship's hull to give stability and the ability to sail against the wind.

LENGTH OVERALL This is the horizontal distance from the fore side of the stem to the aft side of the stern taking in the hull shell but not any protrusions. It should not include the outward reaching length of a bowsprit or stem head fitting, or a bumpkin or mizzen mast boom, etc. (See 'Length over Deck'.)

LENGTH OVER DECK This is a relatively new nomenclature brought in because some yacht brokers have tried to enhance the value of yachts by giving the 'Length Overall' as including bowsprits, etc.

LINES PLAN A set of curved and straight lines which define the shape of a vessel, and are comparable to contour lines.

LOAD WATERLINE The straight horizontal line which runs from where the stem meets the water level to where the stern meets the water when the vessel is loaded.

LUFF FOIL An extrusion made of plastic or aluminium which fits on the forestay and holds the genoa or jib to the forestay. The foil has an aerodynamic shape, and a slot up the aft edge into which the sail front slides.

MIZZEN A mizzen mast is one located towards the aft end of the boat. It is smaller than the main mast. A mizzen sail is set on the mizzen mast, just as a mizzen backstay extends back and down from the top of the mizzen mast, and so on.

PLAN This has two meanings: 1. The view from above looking downwards. 2. A drawing or picture. (See for instance 'Lines Plan' and 'Sail Plan'.)

PULPIT A structure round the fore end of the deck to prevent the crew from falling overboard. The guardrail wires are secured to the pulpit at the fore end and the aft pulpit (sometimes called the 'pushpit') at the aft end. Pulpits are usually made of 25mm (1 inch) stainless steel tubing and have three, four or (less often) six legs.

RIGGING SCREW Also called a turnbuckle or bottlescrew. A tensioning device on the bottom of a wire which supports a mast.

ROLLER FURLING HEADSAIL A genoa which can be rolled up (like a blind) on the forestay. It is popular on cruising yachts because it makes it easy to reduce sail in a rising wind.

SAIL PLAN The drawing which shows the type, size and location of the sails, etc. It usually has details about the spars and rigging.

SECTION A view looking from the bow towards the stern, or vice versa. A 'section view' can also be any view of an object in way of a notional 'knife cut' down, across, or along, to show the inside of the object.

SHEER, OR SHEERLINE The edge of the deck. Also refers to the sweeping curve which the deck follows when seen from the side.

SHEET A rope for adjusting a sail so that it takes power from the wind.

SHEET WINCH A winch which usually has a vertical axle and is located on the deck, or cockpit coaming or the cabin top. It is to give additional power for tightening in a rope and allows one person to do what formerly needed several people before winches became popular.

SHROUDS Wires which support a mast, usually made of stainless steel, with 19 individual strands twisted up into one wire.

SPINNAKER A sail used downwind, cut full like a parachute, and often made of coloured cloth, usually nylon.

STANCHION One of a row of vertical metal tubes secured to the deck. They form a fence round the deck as they support the guardrail wires. Stanchions are normally made of aluminium or stainless steel and located 2 metres

(6.5 feet) apart.

STEM	The most forward part of a vessel's bow extending from the keel.
STERN	The back of a vessel.
STORM JIB	A jib made of extra strong cloth with three rows of stitching along each seam. Used in gale conditions.
TACK	This has two meanings: 1. The forward bottom corner of a sail. 2. The process of turning a yacht when going against the wind so as to bring the wind from one side to the other. It is the manoeuvre which takes the yacht from a 'zig' to a 'zag' as she zig-zags against the wind.
TRANSOM	The flat surface forming a vessel's stern.
TRANSOM STERN	A flat stern which extends to the waterline.
WATERLINE	The level normally reached by the water on the side of a vessel.
WINDAGE	The area exposed to the wind and alternatively the force of the wind on any object. When a yacht is sailing against the wind, windage holds her back and so is an adverse factor.

BIBLIOGRAPHY

The brief notes explain the contents. Some of these books are now out-of-print but can often be found second-hand online, or borrowed from libraries. The ones in print at the time of writing are underlined.

Aero-Hydrodynamics of Sailing
Marchaj, C.A.
The great theoretical book... masses of maths

Basic Naval Architecture
Barnaby, K.C.
Covers a large range of design problems for craft of all sizes.

Boats With An Open Mind
Bolger, Philip
This and other books by this author explain how designers think.

The Boat Data Book
Nicolson, Ian
An essential book for designers with data on proportions, powering, spars, ropes, fastenings, materials, ground tackle... almost everything.

Build Your Own Boat
Nicolson, Ian
Practical information which is a help with construction and sail plans.

Cold-Moulded and Strip-Planked Wood Boatbuilding
Nicolson, Ian
A guide to modern wood boatbuilding for designers, builders and owners.

Comfort in the Cruising Yacht
Nicolson, Ian
A 'problem solver' for designers, owners and builders.

Designed to Win
Marshall, Roger
Deck layout and similar practical design work.

The Design of Sailing Yachts
Gutelle, Pierre
Strong on theory, with lots of graphs.

Metal Corrosion in Boats
Warren, Nigel
The title exactly sums up this important book.

Modern Boat Building: Materials and Methods
Sleight, Steve
Useful for construction planning.

Naval Architecture of Planing Hulls
Lord, Lindsay
Easy-to-understand book about fast power yachts which, though dated, still has many fans.

The Proper Yacht
Beiser, Arthur
A mass of practical information, from an owner's point of view.

Seaworthiness. The Forgotten Factor
Marchaj, C.A.
An important book, full of maths and practical design lessons.

Skene's Elements of Yacht Design
Kinney, Francis S.
The latest edition of what has always been a basic book on the subject.

Small Steel Craft
Nicolson, Ian
Standard textbook on small steel craft.

Surveying Small Craft: Fault Finding in Boats
Nicolson, Ian
A list of designers' mistakes and their consequences.

Surveying Metal Craft
Surveying Sails
Surveying Wood Craft
Nicolson, Ian
These three books are published by the International Institute of Marine Surveying (iims.org.uk) and are all in print. They are fully illustrated and have a lot of design data in them.

Yacht Designer's Sketchbook
Nicolson, Ian
Easily understood sketches of fittings and components.

SKIPPER'S LIBRARY

A SERIES OF PRACTICAL BOOKS FOR THE CRUISING SAILOR

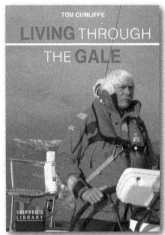

**Living Through
The Gale**
Tom Cunliffe

**Mastering Navigation
At Sea**
Paul Boisser

Weather At Sea
Simon Rowell

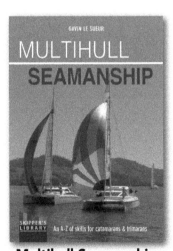

Multihull Seamanship
Gavin Le Sueur

Short-Handed Sailing
Alastair Buchan

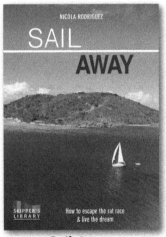

Sail Away
Nicola Rodriguez